ENERGY DREAMS

ENERGY DREAMS

OF ACTUALITY

MICHAEL MARDER

Columbia University Press

New York

Columbia University Press
Publishers Since 1893
New York   Chichester, West Sussex
cup.columbia.edu

Copyright © 2017 Columbia University Press

Library of Congress Cataloging-in-Publication Data
Names: Marder, Michael, 1980- author.
Title: Energy dreams : of actuality / Michael Marder.
Description: New York : Columbia University Press, 2017. |
Includes bibliographical references and index.
Identifiers: LCCN 2016035594| ISBN 9780231180580 (cloth : alk. paper) |
ISBN 9780231180597 (pbk. : alk. paper) | ISBN 9780231542838 (e-book)
Subjects: LCSH: Psychic energy (Psychoanalysis) | Power (Social sciences) |
Political science—Philosophy.
Classification: LCC BF175.5.P72 M37 2017 | DDC 118—dc23
LC record available at https://lccn.loc.gov/2016035594

Columbia University Press books are printed on permanent
and durable acid-free paper.
Printed in the United States of America

Jacket design: Milenda Nan Ok Lee

FOR PATRÍCIA,
WITH THE ENERGY OF LOVE

"L'AMOR CHE MOVE IL SOLE
E L'ALTRE STELLE"

—DANTE, *PARADISO* XXXIII, 145

CONTENTS

OPENING WORDS

Much of my work in political philosophy has been preoccupied with the existential energy boiling under or extinguished in the structures of the state, among other institutions, and in the conduct of "informal" political actors. In *Pyropolitics* I isolated the element of fire that supplied the energy necessary to the conflagrations of revolt and to the calmer, steadier, more controlled flames housed in various "political kitchens."[1] As I developed this line of thinking, it quickly became apparent that a fiery constitution of reality, rather than being limited to a single sphere of human activity, applied to our epoch as a whole. The world is burning and, in its blazing finitude, is reducing itself to smoldering ashes. Postmodern nihilism, global climate change, fracking and oil extraction from the ocean floor, the triumph of technocracy and the "cooling" of the political will—all these are the ramifications of the nearly extinguished pyrological blueprint of reality first furnished by metaphysics eons ago.

The ink had not yet dried on the closing lines of that book and a new and fairly urgent task presented itself, namely to galvanize an alternative, nonviolent framework for thinking about and practically relating to energy without destroying living beings and our planet through its extraction. The scope of such a project could no longer be circumscribed by politics, however broad its conception; it had to be expanded so as to account for our ontological presuppositions, theological aspirations, economic pursuits, psychological self-conceptions, and scientific worldviews. I sought the foundations for an unconventional energy model in the realm of vegetal life, with which I am intimately familiar. As I had noticed in my previous investigations, the energy plants derive from the sun in the process of photosynthesis is nondestructive, world-preserving, and essentially superficial. Could it be that our stubborn denigration of all things vegetal was in collusion with the desire to burn everything and everyone, instead of receiving the bountiful energy of the solar blaze in the manner of vegetation? If so, then we must learn from plants how to live a more ethical life, respectful of the others' claim to existence and operating with a drastically different energy than the one we are accustomed to. The locution *green energy* would need to acquire a literal sense to be truly meaningful.

Although vegetal life winds as a guiding thread through *Energy Dreams*, it is only one entry point into the nonviolent paradigm I seek. Other heterogeneous filaments tied with it in the same knot include Aristotle's notion of *energeia* as the fullness of actuality, later on taken up by Hegel in his conception of *Wirklichkeit*; the hesychastic spiritual practice of stillness that resonates with ashtanga yoga; nonproductivist accounts of divine creation and human work bordering on play; inoperative communities; object-cathexes and what lies beyond the pleasure principle; powerless power and a certain version of perpetual peace; the rescue of matter by quantum physics from its traditional role as a passive substratum for form and action . . .

Inexorably, the thinking of energy will take us back to the roots of Western philosophy, the birth of the concept in Aristotle and its permutations in the millennia after him. Even now, in the twenty-first century, whenever we say the word, we speak in Greek, but its early connotations have become Greek to us. We associate energy with something to be burned, hoarded, or wasted without any clear end, indispensable yet also unidentifiable except by enumerating the resources that contain it. My suggestion is to search for clues to our current confusion and for alternatives to burning the world in the deep past of the concept and the thing, a past that may turn out to be the most radical (and the only) future a living planet can have, namely that of energetic rest, of energy *as* rest and accomplishment.

It is my hope that, in the course of perusing this book, the reader will experience a visceral need, a thirst or a hunger for another energy, irreconcilable with the destructive-extractive procurement of potentiality, power, or force lacking an inherent end. I invite those who pursue not only alternative sources of energy but, before all else, alternative energy as such, to dream with me about its advent. And let us not hurry to label these dreams *utopian* in the face of the harsh reality surrounding us today. We will have plenty of occasions to decide what falls on the side of actuality: the dominant ideology, bent on extracting the last drop of energy from everything and everyone, or the dream of energetic existence that cares for and preserves both beings and being itself.

ENERGY DREAMS

1

ENERGY DREAMS

E *nergy Dreams*—the title came to me all of a sudden, as they say "out of the blue," when I least expected it. It surprised me and, just as swiftly, energized my thought and swathed me in its opacities.

Who dreams, and about what, when energy dreams? Is *energy* the subject here? Or the desired object of a fantasy? Is *dreams* a noun in the plural? Or a verb in present tense, third-person singular? Or, perhaps, both at once? Does energy dream in us, as us, through us? Does it, by so sweeping us off our feet and into its vortex, promote its own increase, its insatiable growth? Is it horrified, if not paralyzed, by the sense of its dwindling? Is it forgetting that, regardless of its peregrinations, it will be conserved, in accordance with the first law of thermodynamics? Or is its reverie one of fullness, completion, and accomplishment outside the instrumental rationality of means-and-ends, which has mutated into the logic of means-*as*-ends?

These are not idle questions that personify a nonhuman concept, now replete with a strangely subjective figuration. Resonating in them is *the crisis of energy* (which is not the same thing as an *energy crisis*), more serious still than the energy worries that have been a part of our lexicon and daily life at least since the 1970s. A salient grammatical expression of the crisis is the equivocation between the verb and the noun we have witnessed in the title of this chapter and of the book as a whole. By force of habit, we think of energy as a resource—a thought not so outlandish considering that, as a word, it is a substantive. A noun, an object, a cause for wars and diplomatic alliances, something to divide, extract, lay claim to, possess. So irresistibly seductive is the grammatical and ontological substantivization of energy that it undercuts our appreciation of its meaning: we begin with different types of fuel or "power" (carbon and oil, solar and hydro) and generalize until we reach a poorly understood and reliably unquestioned umbrella term. The effects of energy, however, surpass a strife-ridden or consensual division of resources. Far from a mere object to be appropriated, it energizes us—our bodies, psyches, economies, technologies, political systems . . . Its sense, then, is evenly split between substantive and verbal significations. The will to energy is none other than the will to willing, where the object, the objective, is not some inert material but an active, activating event—that of the subject. The crisis of energy is that, though treated as a finite resource to be seized in a mad race with others who also desire it, *it* seizes both "us" and "them," taking, first and foremost, our fantasies and our dreams hostage.

Let us isolate two precipitating factors behind this crisis: the relative and absolute ambiguities of *energy*. Relative ambiguity ensues when something is not only unthought but also obdurately resistant to the questioning drive. That is the historical predicament of energy today. Desired in a wholly unconscious manner, dreamed up, even if we keep discoursing and strategizing about it in waking life, it has become little more than a blank screen onto which to project our fantasies of planetary destruction or salvation, enrichment and security, shortages and excesses. Consequently, whatever we say about energy says more about us than about it. Absolute ambiguity, in turn, has to do with the meaning of the concept, incredibly resistant to a univocal determination, unclarified *and*—to a certain extent—unclarifiable.[1] Preceding the wedge modernity drives between activity and passivity, or subjects and objects, energy breaks out and through every frame we wish to impose upon it. So much so that it is, itself, a term in crisis, divided against itself between its current and ancient significations, its ontic clarity and ontological obscurity, its economic desirability and philosophical marginalization.

The absolute ambiguity of energy is an opportunity, rather than an obstacle, for thinking. The situation I have only started to outline reveals that before putting anything or anyone in motion, before releasing heat or the explosive potentialities of things, energy will effect a certain doubling. It will split the atom of meaning in a semantic sort of nuclear fission. This splitting is also happening at this very moment. As I am writing these lines, I am working "on" the thing itself and its concept, but I also cannot avoid paying attention, inadequate as it may be, to the work its chemical, kinetic, mental, and other types exert on me, activating me. After all, the human pleating of consciousness into consciousness and self-consciousness, our attention to ourselves-attending-to-the-world, the whole schizophrenia of humanization is but one of energy's more sublime permutations. Such rifts in what is seemingly unitary (a hobbyhorse of deconstruction) are unavoidable. They fuel every energy dream, including the quenchless desire for its stringent definition and assured possession.

<hr />

Of Greek provenance, the word *energy* is stamped by a double entendre. Composed of the prefix *en-* and the noun *ergon*, *energeia* can be literally translated as "enworkment," putting-to-work, activation. Moreover, the *work* in its midst is nowhere near transparent; we have to work at it, at this work, if we are to appreciate its many nuances.

The range of what ergon signifies is quite broad: from *task* to *function* and from *work* to its *product*.[2] The word repels our ventures to hem it in within manageable confines. It does not keep the distance between the trajectory of a project (work as a process, a task to be fulfilled) and its destination (the function discharged, a product made). With regard to ergon, we are at a loss when it comes to deciding whether we are on our way or have already arrived. In English, we get a taste of this uncertainty when speaking of work: *a* work (say, of art) or *to* work, to produce, to bring to fulfillment. Our relation to energy is fraught and befuddled in part thanks to the plurivocity of ergon, which has in the meantime migrated to other languages, and as far as Japanese with its borrowing *enerugi*.[3]

How did what the Greeks launched or put to work in their idiom drift to other linguistic realities? In what shape has it been received? Has something launched from such a distant time and place really ever arrived? Has it reached us? Is *work*, still unqualified as to its status as a verb or a noun, a singular intimation of the Greek linguistic investment into "energy," which everyone is eager to reap on a global scale?

Consider two alternatives. If the work of enworkment is a process, then energy refers to activation. It sets to work, presumably by interrupting a period of rest, and is itself at work. If it is an outcome, a product of work, then it evinces what happens in actuality, in existence. This second energy is synonymous with the state of affairs, with whatever is the case, the thing itself. We are conversant with the distinction when we classify energy as "stored" or "released." A bomb contains its explosive potential while it is kept in a military warehouse or transported, and it releases its deadly force when detonated. An apple stores the solar energy it imbibed while ripening, but as you bite into it, energy is liberated from its molecules "at rest" (though they are never actually static), counting toward your caloric intake. To us, then, it appears that the divagation from one modality of energy to another is only a matter of time. That which is stored is not yet released, and that which is released is not stored but morphs into another state.

A strange conclusion ensues: energy absorbs time. It does not come about in time; rather, time is activated, or temporalized, in the transitions from one kind of energy into another. Physics corroborates our conclusion through the law of the conservation of energy, which, immune to destruction as much as to being-generated, is merely converted into other forms. Prior to time itself, energy thus veers toward a pantheon of classical metaphysical conceptions and partakes of the *dream of indestructibility*. In Martin Heidegger's thought it finds its place in the illustrious line of misnomers for being, among the Platonic Ideas, "*actus, perceptio*, actuality, representation . . . gathered together in the will to willing."[4] Jacques Derrida, in his wake, tacks it onto the list of "names related to fundamentals, to principles, or to the center [that] have always designated an invariable presence—*eidos, arché, telos, energeia, ousia* (essence, existence, substance, subject), *aletheia*, transcendentality, consciousness, God, man, and so forth."[5] Seeing that energy enwraps the subject and the object, the copula invariably articulates it with itself and paves the way for a tautology at the basis of ontology: energy *is* energy; the work is at work. (One of Aristotle's minimal definitions we will review here strongly resembles this formulation.) In an abbreviated form, the fundamental assertion will profess: energy *is!* Which is to say that nothing is but energy and its enigmatic play, work, or dance around the copula distended into a totalizing movement.

After deconstruction, our theme—wherein we ourselves are ensnared—is understood as a tainted, culpable concept, a dirty word of philosophy, too scientific, too metaphysical, or too economist for our sensibilities. Such stigmatization is inexcusable. The desistance from energy at the theoretical level silently sanctions the most ecologically detrimental methods of procuring it. At any rate, we can proscribe it in thought only by way of its simplification, at the price of its undecidability, its crises and doublings. It is too soon to determine the fate of energy because it is still moving us, we are moved and seized by it, all the while doing our best to seize it under the umbrella of "resources." And it is also too late to determine its fate because the objectified depositories of energy have long become unmanageable and are now threatening if not to annihilate, then to deactivate, to put out of work, out of actuality, the world as such.

Much speaks against a harsh and sweeping judgment that energy, even in the original Aristotelian elocution of energeia, is metaphysical, and so rotten to the core. That is one more energy dream, the chimera of putting it out of action, deactivating it with the help of a relatively straightforward association, by relegating it to the bygone history of metaphysics, the defunct realm of "essential" being. In the text that follows I advance the thesis that the notion, experience, and—if I may put it so—self-experience of energy is infinitely more variegated and conflicted than Heidegger and Derrida concede. Instead of soaking in the stagnant waters of the same, energy is a matter of difference, of transit, transition and alteration, of alterity in being and becoming. A thoroughly homogeneous field would be that of entropy, of energy's divestment, or at least of equalization, where there are no differences between quanta of force, no tension, no life.

In our frenzied activities, we are fleeing from the encroaching shadow of entropic homogeneity, which is why we cling to energy resources so desperately in our personal, national, and globalized existences. The fear of entropy is so intense as to blind us to the kinds of energy we crave, the environmental harm caused by their extraction and burning, the adverse health effects of consuming beverages laced with excessive sugar and caffeine (the so-called energy drinks). The dread of energy starvation, of the looming entropy of reality, pushes us toward what we dread. In response to these fears, which suffuse thought and everyday action alike, one cannot simply denounce the prevailing energy dreams for being the toxic by-products of Western metaphysics, aspiring to an eternal activity, life everlasting, a never-ending erection. Myths do not magically melt away immediately after they are spotted and named as what they are. As far as energy is concerned, we cannot stop dreaming of it, and it cannot cease dreaming us. All we can do is learn how to dream it up otherwise, with our eyes open, knowing ourselves dreaming.

More than anyone else, Aristotle is careful to avoid determining energy, the word he invented or dreamed up, through the apparatus of philosophical definitions. Neither sloppy nor evasive, this theoretical decision gives the thing itself its due, respecting its indeterminacy and singularity. At most, Aristotle offers examples and delineates the term negatively, by contrast to what it is not. In book 9 of *Metaphysics* he breaks his earlier promise to define energeia and suddenly concedes that "we must not seek a definition for everything [οὐ δεῖ παντὸς ὅρον ζητεῖν], but rather comprehend the analogy" (1048a, 35–36). The examples of building and seeing follow (and we will track them shortly). For now, Aristotle defines energy by declining to define it; he substitutes sundry analogies for it in the manner his teacher, Plato, talked of the Idea of the Good by analogy with the supreme and egalitarian dispensation of solar energy. Insofar as Aristotle refrains from defining energeia, in which Heidegger and Derrida recognize his word for being, he resists the urge to behave toward it as if it were an object, a philosophical resource, boundlessly fertile and ready to be tapped into. Doing so, definitively and categorically determining it, would be interrupting its activation, stopping its proper movement in its tracks. But his reticence does not prevent him from saying something (indeed, a great deal) about energy.

Energeia, Aristotle states, "means the presence of the thing, not in the sense which we mean by 'potentially' [ἔστι δὴ ἐνέργεια τὸ ὑπάρχειν τὸ πρᾶγμα μὴ οὕτως ὥσπερ λέγομεν δυνάμει]" (Met. 1048a, 31–32). Plainly, he leans toward qualifying energeia and ergon in terms of actuality, rather than activation, a qualification that pins the work and the at-work-ness of presence on what is not a potentiality, not *dunamis*. Immediately we see that our conception of energy, qua a potentiality waiting to be unleashed into a wide spectrum of activities, is the inverse of Aristotle's. For us, energy is, precisely, not actuality, unless we are sufficiently sophisticated to detect in what presently exists the storehouses of a yet unreleased force. Is this a mere "inversion" of the Greeks? Is our world the Greek universe upside-down? Or, more intriguingly, is there not only a logical-semantic but also a historical, epochal break in the meaning of energy? Didn't the premodern ethos correlate it largely with the accomplished work (the actual, the actualized), even as the modern attitude privileged the moment of a work-in-progress (activation)? If so, then energy's double entendre is not a simultaneous enunciation of more than one meaning but something that takes millennia to work itself out in historical "actuality." Only now, when the livable world is on the brink of collapse, is the concept coming back into its own (albeit not knowing it), in that it connotes the always incomplete activation overwriting the earlier signification of assuagement in the actualized.

Appealing to the values of philosophical rigor and consistency, we might accuse Aristotle of narrowing down the scope of energy in flagrant disregard of his own pledge not to define it (which, in turn, breaks the previous promise to provide a definition). If so, then his energeia is disappointingly one-sided, almost static, boringly present. But this assessment is hardly fair. Being is ceaselessly guided from potentialities to actualities, as "everything changes from what it is potentially to what it is actually [μεταβάλλει πᾶν ἐκ τοῦ δυνάμει ὄντος εἰς τὸ ἐνεργείᾳ ὄν]" (1069b, 15). What is the energy of this change from what is not to what is energeia (being)? How can something issue from nothing, specifically in ancient Greek thought? Taking care not to dispense prematurely with the disquietude of these queries, note that the other pole, against which Aristotle first negatively defined energy, is not nothing but another ontological condition—dunamis or "what is potentially." (Is the state of this *is* itself potential or actual? That will be the gist of Plotinus's inquiry in *Ennead* II.5.) Needless to say, when we mouth the words *actuality* and *potentiality*, we do so in Latin, not in Greek, though, in Latin they also say more than what we habitually glean from them. Potentiality derives from *potestas*, which, as one possibility for translating *dunamis*, signifies *power* and *capacity*. Strangely, then, the global movement from potentiality to actuality implies a diminution of the ontological power-capacity, its desaturation. Insofar as a being is actualized, or energized, it fulfills itself. Less and less does it require the enabling power of dunamis to attain the condition that corresponds to it according to what it is. τὸ ἐνεργείᾳ ὄν, what it is actually (say, a fully grown oak tree, rather than an acorn), is the outcome of the work performed by the essence of the being it is (here: oakness) in the world. It is and therefore it has no need of the power-capacity to become. Its past workings rejoin the ensuing work in energetic plenitude. Aristotle's energy is thus markedly powerless and incapable, not in the sense of lacking these qualities, not by way of deprivation, but because it is beyond the vicissitudes of dunamis, which names an incompletion to be overcome. In its powerlessness it is possible to detect a nonmetaphysical trail through the proverbial thickets of metaphysics.[6]

We are still circling around the vagueness of energeia in Aristotle, who qualifies the makeshift contrast he has established by allowing for degrees of actuality and hence of being. If a fully grown oak tree still keeps on growing, that is due to the necessary incompleteness of its accomplishment, imperfect (*atelos*) in its very idea, a haunting absence persisting in its presence, calling for more power-capacity to be discharged. Although Aristotle thinks that vegetal life is exceptionally prone to this predicament, every actuality is partial and potential vis-à-vis the next actuality, "since no action which has a limit is an end but is only a means to an end [ἐπεὶ δὲ τῶν πράξεων ὧν ἔστι πέρας οὐδεμία τέλος ἀλλὰ τῶν περὶ τὸ τέλος]" (1048b, 18–19). Finitude is diffracted into a limit and an end, *peras* and *telos*, and it is in the caesura between the two that confusion between actuality and potentiality ensues. The absolutely actual is the end, whereas limits admit of nothing but deficient actuality mixed with potentiality. With this pretext, Aristotle will conjure up the most famed and influential energy dream in the history of Western thought, the dream of the unmoved mover.

What becomes of energy in our world of constantly surpassed limits devoid of ends? In this world of sheer means, the end is brutally imposed from the outside as the limit to all limits, a term, terminus, or termination that, in the withdrawal of energeia, breathes with destitution, nonfruition, the void, an arbitrary cut. For us, the end is singular, irreplaceable, final; it is the only unsurpassable limit, that of death. Heidegger's *Sein zum Tode*, being-toward-death, occupies the vacant place of an end, the telos, shorn of an inner purpose and of a subject who would enjoy its actualization. It motivates, energizes our thoughts and actions, including those that attempt to evade it, and yet does not bring human existence to fruition in the fullness of energeia. An entire classification of ends, promising assorted energy streams, trickles, and dams, awaits us.[7]

On a macroscale, the end of the planetary world mirrors the end of the world that each of us is. The race to extract energy from everywhere and everything, whether by drilling for oil in the Arctic seas or by distilling biofuel out of living plants, is instigated by a growing sense of nonaccomplishment, which this race actually aggravates. So long as something still persists in actuality, it is taken as an invitation to a work yet to be carried out, the suicidal work of separating and releasing energy from matter and dissolving the temporarily stabilized structures of our phenomenological lifeworld into dynamic processes. The limit to all limits and the limitless quest for energy are locked in a virulent dialectic: irrespective of our efforts to defer the a-teleological end of the world—indeed, thanks to these very efforts—we hasten its approach. A thorough determination of *what is* is tantamount to the termination of existence through the final release of energy from matter, and an absurd release at that. Not actualizing anything whatsoever, the endless end, the activation of a diabolical metabolism digesting actuality into potency, reverses the river of being that (with the exception of the unmoved mover) flows from dunamis to energeia.

We are so enthralled with the possible that we've ceased caring about the actual and forfeited the energy of the latter. For us, actuality is but a way station on the highway of unlimited possibility. We draw energy from the assumption that potentiality is limitless and that this limitlessness makes it possible to begin with. What is behind such suppositions, however, is not only the neglect but also the destruction of actuality. We render finitude all the more precarious and finite the more we speak and act in the name of infinite possibilities. We procure energy from the destitution of energy suffusing and dissolving the fabric of *what is.* The greatest enemy of contemporary humanity, of our planet, and of material existence as such is thus unbridled possibility that more and more renders the world impossible.

Even so, there are activities that, their finitude notwithstanding, contain their ends within themselves. Vision, for one, is not a gradual shift from the potential to the actual; its energy is such that it is at every moment actualized in a mélange of tenses. "One has seen as soon as one sees [ὁρᾷ ἅμα καὶ ἑώρακε]," Aristotle observes, but one has not built as soon as one embarks on the project of building (1048b, 23). Vision shares this energy with understanding and thought ("one has understood as soon as one understands"; "one has thought as soon as one thinks") and, more important, with a good life and with happiness. As soon as one lives well, one has lived well and as soon as one is happy one has been happy: "εὖ ζῇ καὶ εὖ ἔζηκεν ἅμα, καὶ εὐδαιμονεῖ καὶ εὐδαιμόνηκεν" (1048b, 25–26). The energies of vision, understanding, thought, a good life, and happiness absorb time, while remaining faithful to the temporal reality of "this world." Energeia, Aristotle intimates to us in what will be taken as an untenable position, comes into its own in conditions that obviate bustling activity and vigorous pursuits. It resides in the quiet of visual or thoughtful contemplation, in the joy of understanding, in the radiance of happiness outside time in time.

We have been conditioned (in part by the history of metaphysics and theology, in part by the voguish dogma that prides itself on resisting the tradition it inverts) to visualize no more than two alternatives: either the ideal of completion, unattainable in finite existence, or the reality of incompletion, congruent with finitude: either the plenitude of energy that does not change or a chronic lack of energy (and of time) that diminishes along the course of our biological lives. Hidden from us is the third option of inexhaustible energy in every instant that fleets by solely according to the perspective of those who are, themselves, rushing without noticing its inimitable singularity. Letting go of this boundless source of wonder and inspiration, we bar ourselves from the shimmer of happiness, which may not be so different from the brilliant givenness and reciprocal exposure of the world and our senses, with all the energy transmissions that happen between them. Who still has the courage to be adrift in the virtuous circle of "one has seen as soon as one sees," where the present is perfect and perfected, grammatically and experientially? Or, to abandon oneself—which is here the same as finding oneself—in the labyrinths of happiness, where one has been as soon as one is?

It is questionable, under these circumstances, whether we can still fathom the Aristotelian energy of contemplation and happiness where being fully at work, activated or actualized, sparks the experience of rest-fulness beyond the opposition of activity and passivity or beyond the metaphysical contrast between the temporal and the eternal. Movement and rest are comparable to two parallel lines that, in open vistas, appear to intersect at an infinitely removed point on the horizon. Conversely, in our humdrum reality, time slips away, passes; energy ebbs away; we unsuccessfully try to catch up with both and lose ourselves along the way. Work and rest are mutually exclusive, and the leisure of resting, connoting laziness and indolence, becomes more and more rare. We are chronically short of the time to see or to think, to be happy or to live a good life, for that matter. One has seen *before* one sees, has understood *before* one understands; the world comes to us preinterpreted in the shape of old news, anticipated, worn out, printed in a newspaper long after it was announced online or spread as a digital rumor. Hermeneu-tics has taken advantage of this predicament as its linchpin ("the 'world' which has already been understood comes to be interpreted"),[8] while omitting that something of Aristotle's energeia is indispensible to the progress from preunderstanding to interpretation. We feel the crisis of energy only very obscurely, through the prism of lack, a dearth of time and resources, for which we compensate with the acceleration of restless quests for more means bereft of an end and experienced, in a moment of involuntary "downtime," as the depletion of ourselves. Perhaps we dream of rest, of deceleration, of abiding in and with plenitude, but it is just that—a dream. "And it's an eternal struggle! Rest is only the stuff our dreams . . . ," my grandmother was fond of saying in what I later discovered to be a citation from a poem by Alexander Blok titled "On the Kulikovo Battlefield."

———— ❧ ————

Throwing the entire weight of his thought against the lurid spectacle that is our actuality emptied of the actual, Aristotle dreams up pure energy dreaming itself. And in such a way that it would not be a fantasy but the most palpable and secure reality! Aristotle thus dreams of energeia in the name of the most real, least dreamy state, conjuring up, along the way, the neologism energeia that is his singular invention.

As a safeguard against the infinite regress of a merely potential actuality, he asserts that "there must be a principle of this kind whose essence is actuality [ἄρα εἶναι ἀρχὴν τοιαύτην ἧς ἡ οὐσία ἐνέργεια]" (1071b, 20). In the beginning, in principle (arché), there will have been pure energeia that did not require further actualization and did not spring from a preexistent potentiality. Yet Aristotle is also cautious. He keeps vigil over his reverie so that it would not degenerate into a pipe dream. The Greek thinker identifies an aporia (καίτοι ἀπορία: translated as "difficulty") in this postulate, which contravenes our empirical common sense: "for it seems that everything that actually functions has a potentiality, whereas not everything that has a potentiality actually functions; so that potentiality is prior [δοκεῖ γὰρ τὸ μὲν ἐνεργοῦν πᾶν δύνασθαι τὸ δὲ δυνάμενον οὐ πᾶν ἐνεργεῖν, ὥστε πρότερον εἶναι τὴν δύναμιν]" (1071b, 22–25). There is always an admixture of dunamis in everything actually in existence, and, vice versa, dunamis can perfectly subsist as it is (or, better, as it is *and* is not) without being transmogrified into energeia. How to tackle this aporia? How to pass through it while preserving the primacy of actuality over potentiality, the primacy that, inspired by the philosophy of F. W. J. Schelling, Heidegger will invert or subvert in the opening paragraphs of *Being and Time*?

Aristotle's ratiocination in book 12 of *Metaphysics* deserves careful study and attention. His solution, the "unmoved mover," is not a theoretical trick, a magician's "rabbit-out-of-a-hat," meant to hide the aporia behind a philosophical smokescreen. It commences with a deduction from the empirical observation that potentiality did not stay locked in its impoverished mode of being but advanced toward actualization, if only partially and sporadically. Aristotle makes the actual, the work, ἔργον, testify to this premise. Since it has come out of itself, "there is also something that moves it. And since that which is moved while it moves is intermediate, there is something which moves without being moved; something eternal which is both substance and actuality [ἔστι τι ὃ οὐ κινούμενον κινεῖ, ἀΐδιον καὶ οὐσία καὶ ἐνέργεια οὖσα]" (1072a, 20–25). Forget the chicken and the egg! The rabbit appears from the hat as much as the hat from the rabbit, while the difference between the two qua actualities fades away. More than that, Aristotle would have been prepared to accept the tenet of infinite regress, if existence were comprised of abstract nonactualizable potentialities. However, given the evidence of works-facts, *erga*, and beautiful phenomena-shinings, instances of *phenomenon kalon* (1072a, 28) that surround us, he has no other choice but to conclude that being is not dunamis all the way down. The energy of the whole must have originated in actuality fully present to itself that, overfull with itself, spilled over into other kinds of life, mixed with potentiality, and swung closer to limits than to ends.

At this point, readers' patience is likely to run out. How can one vindicate such obviously metaphysical constructs? Are they really preferable to our topsy-turvy conception of energy as dunamis? Do we need to go as far as to resurrect the phantom of an end without (prior or intermediate) means to resist the ruinous force of means devoid of an end?

The unmoved mover is, admittedly, nowhere to be found among phenomena and facts. But this conspicuous absence is not a good reason to attach to it the label *otherworldly* or *metaphysical.* I take the unmoved mover to be a point where energy splits from itself and, in this explosive splitting, relates to itself anew. (Empirical inaccessibility does not automatically spell out metaphysical abstraction. Who has ever seen, touched, tasted, or smelled a relation, let alone a self-relation?) The unmoved mover is the self-relation of energy, or of energies, just as the soul is the body's relation to itself, as Hegel correctly intuited from the Greeks. In this regard, I find it instructive to consult the interpretation of Aristotle by Russian philosopher Alexey Losev, who casts the unmoved mover in terms of being's self-relation.

Schematically, Losev's exposition may be divided into five stages:

1. The world is in perpetual motion, which is its expression and which stands for the *energy of meaning*.
2. Since there is nothing except this energetic meaning, it relates to itself by thinking itself. Hence the energy of the world is also self-thinking, the *energy of the Mind*, thought thinking itself.
3. This mindful energy can give itself varying degrees of material expression, with the highest expression of essence in a singular fact being beauty or perfection.
4. Yet, more often than not, the journey of energy into limitless materiality is a transition from abundance to lack, from happiness to unhappiness, from self-satisfaction to suffering and self-abandon.
5. Although the energy of the Mind forgets itself in its wanderings, it remains gathered into itself in a single point. The Aristotelian doctrine (of the unmoved mover) is the memory it keeps of this self-gathering.[9]

The articulation of energy with itself in (1) and (2) is coterminous with its splitting. In relating to itself, energeia must first become other to and differ from itself. Metaphysically inclined, we are prone to divide the energy of meaning and the energy of the Mind into two separate realms, two worlds shaping the landscapes of physical reality and metaphysics, respectively. And yet it is the same divided against itself and—at times uneasily—reconciled with itself. In those exceptional instances when reconciliation succeeds, beauty and perfection flash before us (3). More frequently, the faulty sutures between the disjointed fragments of energy botch its self-relation, culminating in a tragedy that marks the human condition (4). To make things worse, our work on this tear is futile: on behalf of metaphysical energy dreams—that is to say, of a seamless and extratemporal perfection—we further isolate material meaning from the knowing mind, thereby deepening the cut.

Upon finishing his monumental study of ancient thought in 1928, Losev could not have anticipated the pitch the tragedy has attained today. Un- or disarticulated, the sundered hemispheres of energy are also out of touch each with itself. The energy of meaning mutates into dense meaninglessness, which is how matter presents itself before the scientifically determinist consciousness. Nor is the energy of the mind spared in its current form. With thinking reduced to the bare bones of calculation, there is no longer a question of its self-relation, save for the narrow channel of "criticism" or, rather, "self-critique." As the tension between and within the energies of meaning and the mind slackens, nihilism emerges as the destiny of the world. It could well be that Aristotle is ahead of us, insofar as the energy of plenitude, life, and rest are concerned (5). He is waiting for us to catch up with him. But it could also be that the self-forgetting of energy, as if in a dream, is a more effective recipe for anamnesis than the actual memory of actuality present in Aristotle's texts. After all, the multiple fissures and rifts that traverse energy (the concept and the thing itself) call for something a little more complex: a forgetting that remembers, a dream that is more actual than our dunamis-obsessed virtual reality, a suture that displays the cut, a wounded spirit whose flesh is covered with unhealing scars.

Dunamis is dynamite. That is the unconscious image of energy nestled in our psyches. To obtain it, we must extract it, wrest it from others—human or not, animate or inanimate—at the price of their wholeness. Energy extraction tears actual beings, the placeholders of the Aristotelian energeia, apart, obscenely exposing their entrails, enucleating them. Energy production is a fury of destruction, which Hegel incidentally associated with "universal freedom,"[10] reinforced by the implacable conception of energy itself as something indestructible. It does not relent until the atom is split, until it reaches the nucleus and divides the ostensibly indivisible. Nuclear power and the atomic energy it unbridles is the apotheosis of the contemporary dogma of energy, the fulfillment of a process forbidding any fulfillment, anathema to everything actual. So is hydraulic fracturing, or fracking, that cracks the earth (particularly shale rocks) open by exerting high water pressure on them from below. Environmentally destructive and shockingly shortsighted as these methods of energy production are, they are to be expected given the dominant conception of energy that requires breaching and laying bare the bedrock of things (of the atom, of the earth) so as to draw power from this violent exposure.

The crisis of energy is at its direst here. Enucleating the world, we ourselves are enucleated. Be it labor or truth, we extract value from the universe outside us *and* from the human, destroying the material "shells," the unacknowledged substantive expressions of energy that enclose its active "kernel." On the spreadsheets of capitalism, we are accounted for as *human resources*, from which work can be extracted, burying Marx's dream of human self-actualization through labor. Our epistemologies, too, acquiesce with the ambition to disclose the marrow of reality, usually by shattering and discarding the outward "mere" appearances that occlude it. Thinking has assumed the shape of mental fracking. Unless we subscribe to a phenomenological methodology, we are quite dissatisfied with the surface of things, with how they present themselves to us in everyday life: with their imperfections, incompletions, shadowy spots, and badges of finitude, be they limits or ends. Our readiness to pump oil from the ground or from the ocean floor belongs with the rest of this epistemological apparatus, which is why it is so difficult to give it up. For us, superficial actuality, the actuality of the superficies, is never actual enough. As we strive to know what things really are, we break them down to atomic and subatomic, chemical and molecular components. Why would the style of energy production and extraction be at variance with that of the production and extraction of knowledge? The two would have to change in tandem, if human impact on the world, as well as on ourselves, is to be mitigated. We are yet to register the (energetic) repercussions of Hegel's dictum concerning the rationality of the actual and the actuality of the rational. "What is rational is actual; and what is actual is rational."[11]

On the one hand, most approaches to energy presuppose a substantial divergence between the inner and the outer, depth and surface. The very language of *storage* and *release* indicates that the energy of everything from galaxies to microbes, from economic systems to psychic life is contained, held within, withheld, and prevented from achieving its full actuality, before it is liberated with more or less force but without end and to no end. The encompassing whole (i.e., the physicists' "closed system") is likewise seen as a great container from which no energy ever escapes; that is what, at bottom, the law of energy's conservation, the first law of thermodynamics, intends. Absent the dimension of interiority, one would no longer be able to explain how things work, how they are put to work, activated, or withheld in potentiality. Energy differentials depend, above all, on the difference between the inside and the outside, on the speed and force with which these boundaries are, or may be, traversed.

Plants, on the other hand, need not devastate the interiority of another being to procure their energy. They set to work the elements they do not control, do not dominate, do not appropriate. Besides water and the minerals they draw from the soil, they receive what they need from the sun, collecting solar sustenance on their maximally exposed surfaces, the leaves. (Plants can, to be sure, deplete the soil, but this results from human interference—the introduction of intensive agriculture and the spread of monocultures. By and large, through decomposition, vegetation returns to the earth much of what it has taken from there.) Human reliance on solar energy would signal our willingness to learn from plants and to accept, *mutatis mutandis*, an essentially superficial existence or, at the very least, to integrate it with the dimension of depth. Although current technical capabilities could sustain a nearly total reliance on renewable energy (solar, wind, hydro . . . : I prefer to designate these not as "renewables" but as "elementals"; the former category makes little sense, in that it lumps together plants grown to be burned as biodiesels and our cooperation or cohabitation with the elements of water, air, and solar fire), although these capabilities exist, they are extraneous to the prevalent mindset surrounding the *essence* of energy, to be encountered in the deepest depths, to be accessed through destructive-extractive means, and to be snatched from the interiority of things. The focus of attention may actually shift to "clean energy," and that is, in and of itself, laudable. But "cleanness" relates primarily to the effects of utilization, not to the question of what energy is. That is why oil, coal, and, most of all, natural gas companies can profess that they are making the transition to clean energy, without radically modifying the sources of fuel themselves, let alone how they are procured.

The most vital energy dreams, then, are the dreams of another energy, nonviolent and enacted otherwise. Long ago, a form of life was dreamed by energy itself that approximated this desideratum. That form was vegetal. Its vital exposure, the essentially superficial attraction to solar energy, is a polar opposite to the mode of thinking and living that values depth. After the technical and theoretical network of energy and knowledge production has permeated and analyzed the core of things, an additional kind of depth and interiority—spiritual or metaphysical—is constructed. Not by chance, plants have been excluded from the province of metaphysical spirit and consigned to its margins. Their life looks alien when memory, consciousness, and the psyche are imagined in the shape of invisible inner drawers for storing experience and for the release of mental energy into specific behaviors. *To vegetate* and *vegetative* have come to denote torpor and a dearth of energy, in short, what human beings ought to avoid in order to lead a vibrant life.

In its "commonsense" permutation, *energy* is highly seductive. The ultimate prize, it beckons with movement and dynamism. As such, it offers the possibility of possibility, a magic wand for fulfilling any desire. To maximize energy is not so much to hoard more of its resources as to liberate more of it, ceaselessly traversing the boundary between the inside and the outside. The ideal of this activity would be to altogether eliminate the time interval between storage and release. Scarcely pleased with stability, modernity is inebriated with perpetual change or, less charitably, with the illusion thereof. If physicists and chemists speak or dream about the *liberation of energy*, what they leave unspoken is the assumption that matter is its prison, a cipher for containment, curtailing dynamism.

We might recall that the work of energy—its *ergon*—is ambiguous to the *nth* degree. It embraces the process and the product, activity and actuality. In physical reality, in political systems, as much as in the psychological sphere, inertia, stagnation, and the status quo are energy factors under the substantive aspect of ergon, and revolutionary, innovative, groundbreaking actions unfold under ergon's verbal aspect. That energetic systems are heading toward entropy is an assertion disregarding this basic equivocality. (The Greek *stasis* is similarly amphibological in its meaning: it pertains as much to immobility, or stagnation, as to the intense strife of a civil war.) Now, since in our collective unconscious energy is untethered to the logic of ends, reverses direction from actuality to potentiality, and operates within the stricture of destructive extraction, it sows death not as a result of grinding to a halt but by virtue of an exorbitant movement that discards every temporarily stabilized form. Our energy unfurls the force of negativity, actualized not in being but in the decimation of being.

Is Hegel complicit in this dreadful scenario? You might get the impression that you have made out the contours of his thought on the horizon of energy as I have sketched it out thus far. It is true that dialectics is a machine of energy production-extraction from rational principles, with negativity (more precisely: self-negation) for an internal combustion engine. And yet Hegel kept too near to Aristotle to endorse the "bad infinity" of unalloyed potentiality. What he extracts from the deep and essentially hidden rational kernel are the actual structures of physical, social, and political reality. "What is rational is actual [*wirklich*]; and what is actual is rational" is his manner of underwriting the Aristotelian energeia, translated into the Latin *actus* or *actualitas* and, through it, into the German word for actuality, *Wirklichkeit*. The extraction of the actual from the rational is not a lethal operation. In self-negation and the negation of the negation (inspired or energized by Christian resurrection) death is incorporated, overcome, and put in the service of actuality. The derivation of dialectical energy is the work of history, as the history of Spirit, the self-relationality of the world's self-relation.

Nor does Hegel lose sight of the powerlessness distinguishing the Aristotelian energeia. The preface to *Philosophy of Right*, where he pithily articulates the actual and the rational, holds another well-known statement comparing dialectical philosophy to the Owl of Minerva that "spreads its wings only with the falling of the dusk."[12] The time of dusk is when the rational is *almost* actual, when it is too late to change anything, when one can but observe and recapitulate in thought that which has been consummated in the world. When nocturnal dreams, oddly in sync with total alertness, are encroaching upon us in the twilight of an immediately living actuality that marks the transition to Spirit's absolute Wirklichkeit. "The restlessness of the negative" aside, dialectics is the energetic rest of fulfillment, in the absolute achievement of the absolute, hastily dismissed or feted as "the end of history." It is thought thinking itself in the form of rational actuality, revisiting *what is* as it is (ergon as a work) and going over the past process of its coming-to-be (ergon as working-out or setting-to-work) replayed on the record of the absolute.

In the end, after the end of immediate actuality, dialectical energy bears the self-relational, self-mediated traits of the unmoved mover. Restlessness in absolute repose, where does it fit on the subject-object continuum? While, in *Lectures on the History of Philosophy*, Hegel equates the Aristotelian energeia to "the principle of subjectivity," and dunamis (which he calls *possibility*) to the objective moment,[13] in *The Science of Logic* he implicitly links energeia to the final subject-object synthesis. In the latter work, when "the Idea posits itself as the absolute unity of the pure Notion and its Reality, and thus gathers itself into the immediacy of Being," it experiences "an absolute *liberation* [*Befreiung*], having no further immediate determination which is not equally *posited* and equally Notion."[14] As a result, "the Idea freely releases itself in absolute self-security and self-repose [*die Idee sich selbst frei entläßt, ihrer absolut sicher und in sich ruhend*]."[15] The most crucial liberation of dialectical energy, which frustrates the expectations of a chemist or a physicist, occurs in this release of the Idea into actuality, a discharge that, rather than devastate *what is*, affirms it. It will be objected that the utopian harmony of "pure Notion and its Reality" is only celebrated in the end, when it is too late to salvage "the real world" from the negations and detonations it suffered through. But the dialectical end is, by the same token, the beginning: dialectics commences with or after the realization of the absolute such that "absolute self-security and self-repose" underlie its every torsion and turbulence. Hence, the liaison between the unmoved mover and free energy, released into the *completed* work of negativity.

The surface/depth scheme stays intact only on the condition that we ignore the absolute beginning of dialectics from the absolute. Seen absolutely, actuality is the external existence of the rational in which the previously buried essence manifests itself.[16] That said, according to *The Encyclopedia Logic*, Aristotle's most significant advance over Plato was to spare the Idea the fate of being dunamis and to confer on it the advantages of energeia, "the inwardness that is totally to the fore [*heraus*], so that it is the unity of inward and outward." (I hope that you are starting to discern the silhouette of a growing plant here.) "In other words," Hegel continues, "the Idea should be regarded as Actuality [*Wirklichkeit*] in the emphatic sense that we have given to it here."[17] Energeia is actuality as the synthesis of the inner and the outer, of essential depth and phenomenal surface. It is the essence (or the heart) worn on the sleeve, totally "to the fore," *heraus*. This explains why the liberation of this energy is free of residual violence. A self-actualized self-actualizing, it works on itself, no longer infringing upon the boundaries between interiority and exteriority. Hegel resolves the crisis of energy by determining the term, dismissing its undecidability (not on a whim, but through meticulous mediations), and extinguishing the tension between the two senses of ergon. A *work* and to *work* meld together. Dialectics supplants the immediate doubling of actualization and the actual with the mediated repetition of work after it is done, the inwardness of en-ergeia articulated—spatially and symbolically—in the outside world.

To return to my initial questions: Who dreams of energy? What does energy dream of? Who or what dreams when energy dreams? Hegel gives a univocal response: Spirit, both as subject and as substance. And yet energeia has shown more affinity to "the principle of subjectivity" than to "objective possibility." In what sense, then, does it envelop substance? Proceeding dialectically, we must conclude that the inwardness of subjectivity is modeled on that of energeia whose interiority turns inside out in the course of accomplishing itself. Whoever or whatever is at work or in the work is thereby outside itself in itself, engendering an excess over itself. That is to say: all work is the work of love, as this very engendering. Contemplated from the side of "objective possibility," the subject is the supplement of actuality who does not add anything new to the real. But the absolute has no sides, which is why it makes both possible and actual the concretion of substance from the energetic excess of the subject over itself. Energy dreams (are) about the identity of subject and substance.

The drawback of the Hegelian solution is that it cannot precipitate sorely needed paradigm shifts in knowledge and energy production. True, the Owl of Minerva will take its flight over a devastated earth from which every drop of oil and every cubic meter of natural gas have been extracted. But, rather than fault the theory of energeia and philosophy itself for their enthralling, if irresponsible, powerlessness, we ought to attend to a figuration they are seldom aware of. In the exposure of their essence, "the inwardness totally to the fore," energeia and dialectics remarkably assume vegetal outlines. Hegel must have had a premonition of this metamorphosis when, in the preface to his *Phenomenology*, he likened stages in the life of Spirit to phases in vegetal growth and repro-duction, or when, in *Lectures on the Philosophy of World History*, he wrote: "Spirit is essentially the result of its own activity; its activity is the transcending of immediate, simple, unreflected existence,—the negation of that existence, and the returning into itself. We may compare it with the seed; for with this the plant begins, yet it is also the result of the plant's entire life [*Wir können ihn mit dem Samen vergleichen; denn mit diesem fängt die Pflanze an, aber er ist auch Resultat des ganzen Lebens dersel-ben*]."[18] Activity and its result, the working and the work (on oneself)—the principal moments of energy condense and converge in a seed. Giving itself a determinate energetic form, dialectics elects a vegetal figuration. It dreams of setting itself to work by flourishing, blossoming, coming to fruition, and germinating again in and as Spirit. That is, by endorsing the plant's essential superficiality, derided in *Philosophy of Nature*.

Unjustly, inexcusably, I have let slip the first two letters of *energy* in the midst of my concentration on the polysemy of ergon. Unless I have been dealing—obliquely, not yet thematically—with nothing other than these letters of the prefix *en-* meaning *in*. Of course, the most obvious sense of "in" is that of spatial inclusion, which goes some way toward explaining the language of containment in the treatment of "unreleased" energy. When liberated, however, energy is not released outside some given spa-tial parameters. It transgresses the boundaries of things that temporarily stabilize and detain it, but, in this transgression, it is imparted to another object, the recipient of its impetus, or, if that is not the case, is diffused into space, however vast or narrow.

It follows that "inhood" cannot be reduced to the sense of spatiality; the ideal involvement in the work supersedes the physical interiority of the object. The *en-* of energy in *enworkment* signifies, among other things, being in a process, under way, unfolding. The actual is what is *in actuality*, which also means now, at present, in the present moment, itself as elusive as energy, since every *now* swiftly turns into a *then*. In contrast to spatial stability, the temporality of energy can be fitful and erratic. At work in this moment, it may be deactivated the next. Because we associate time with perpetual movement and change, and because time depends on the tempo or rate of energy conversions, a break between containment and release can connote a disruption in energy's work, its being out of work, decommissioned. That is what we are afraid of—an abeyance of energetic conversions, their work no longer at play. The real exteriority to what is in or at work in energy is temporal rather than spatial.

If we press a little further, we will run aground on the question of time's exteriority. Isn't it as ludicrous as that of the spatial variety? Is a rupture with finite existence (temporal through and through) conceivable in the cessation of the changes it undergoes, a cessation, at any rate, illusory, to the extent that energy keeps fluctuating above and below the thresholds of human perception aided by the latest microscopes and telescopes at our disposal?

These doubts evaporate upon taking the phenomenological experience of time into account. To us, the inheritors of modernity, rest, the withholding of energy, a sense of stability, and a perceived absence of stimulation feel like slowdown and death. A deceleration tending to zero speed might be only an imagined possibility. All the same, it makes us cringe, mobilize all the energy resources we can muster, and engage in a tenacious self-activation intended to defer the final pause of mortality. Being-out-of-work in economic and ontological domains signifies quitting the time of activation, which metonymically stands for time. (In Spanish, *unemployment* can be said in two ways: *desempleo*, a word that has the same Latin-derived root as its English equivalent, and *paro*, "stoppage.")

For the ancients, inversely, a suspension of time in time signified happiness, the fullness of consummation, the most intense energetic state imaginable. Their temporality of *en*workment overflowed the metaphor of time-as-a-flow.

Both temporal and spatial, the twofold inwardness of work in energy reflects the verbal and substantive aspects of ergon. Besides instigating the unfinished process of activation (as the verb), energy is operative in the physical thing that "contains" it (as the noun). It goes on working in the course of bringing something or someone into existence *and* in the resultant work itself. Its containment is never simple, having to contend with pressures, resistances, entropy, and so forth. The spatiality of things is home to the temporality of work that is unabated in them after they have been "produced."[19] We should get accustomed to hearing in the spatial overtones of the word *in* energetic transactions astir alongside the more receptive inclusion and envelopment.

My discussion of the preposition *in* is indebted to the theme of "existential spatiality" Heidegger developed in *Being and Time*. There he insisted that human being-in-the-world is not of the same nature as the "insidedness" of water in a glass (the example is Heidegger's). Apropos *Being-in* [*In-sein*], "we must set forth the ontological constitution of inhood [*Inheit*] itself,"[20] he submitted. Human ontology demands existential inhood, and the ontology of things requires categorial inhood. Energy, for its part, is capacious enough to encompass this Heideggerian distinction. The polysemy of ergon backfires on the meaning of *en*: *a work* and *to work*, it muddles the contrast between the categorial nature of a product and the existentiality of an activity, *as well as* between two types of "the ontological constitution of inhood." The energy of water in a glass or of a wave hitting a rock reflects and is reflected in the energy of human beings moving about in, organizing, and despoiling their world. Between the two—a speculative identity in difference, an infinite reflection of the reflection of the reflection . . .

Heidegger himself must have been nebulously aware of this problem, particularly as he composed his influential essay "The Origin of the Work of Art." He emphasized there that artworks are both works and things; they "universally display a thingly character," but nothing can be said about this character "so long as the pure self-subsistence of the work has not distinctly displayed itself" in them.[21] The very being of an artwork, which is a work and a thing, deconstructs the difference between the existential and categorial analytics jealously guarded in *Being and Time*. While appearing to be an object among others, a work of art (e.g., a Greek temple—again, following Heidegger's own example) is in the world in such a manner that it first opens the world, articulates the elements, gathers them into a coherent whole: "The temple-work, standing there, opens up a world and at the same time sets this world back on earth, which itself only thus emerges as native ground."[22] The work of an artwork is, precisely, this disclosure of the world, spanning the specialized realm of aesthetics and the vast fields of *aesthesis*, of sensitivity, perception, feeling.

And energy? Qua actuality, Wirklichkeit, it presides over the demolition of barriers between existence and categories: "Art is actual in the artwork. Hence we first seek the actuality of the work. [*Wirklich ist die Kunst im Kunstwerk. Deshalb suchen wir zuvor die Wirklichkeit des Werkes.*] In what does it consist? Artworks universally display a thingly character."[23] Art does not spawn a dreamworld. It boasts a reality (thinghood) and an activity (work) in which its actuality-energy inheres. The work works on and takes place in the thing; the thing works by articulating the world by which it is itself articulated. An artwork is *in* the work and *at* work, connecting the categorial and the existential takes on being-in. I am not fond of philosophies that indulge in generalizations from aesthetic practices. But, for all that, it would be warranted to extend Heidegger's handling of the energy of art to other "ontic" realities. This gesture would go a long way toward dissociating being-in from pliant inclusion in an empty, undifferentiated, and infinitely stretched-out spatial milieu.

Whether a process or an outcome, *work* is tinged with economic hues. Energy, by implication, clothes itself in the trappings of a congenitally economic concept. More than that, an economic mentality compresses its work and its workings into the sphere of production. Erga are productive activities or products, means or ends within the same overarching order of quantifiable outcomes. There is no qualitative disparity between the means and the ends within the productivist account of energy, expended to achieve a tangible (or, at least, measurable) result and objectified in that which has come out. Energy here is the power of actualization, the power temporarily kept at bay, productively employed, or preserved in the actuality of its products. Its default conception is incongruous with powerlessness, except for the negative modality of energy: deprivation, shortage, crisis. In the shadow of production, potentiality is only a prospect not yet productively realized, and actuality—a trace of previously activated productivity. Substantively the same, the past and the future of production contribute to the vanishing of the present incommensurate with a product and with its fabrication. Along with the positive powerlessness of energy, time disappears.

The double entendre of the Greek ergon is still with us, albeit stripped of its patent ambiguity. *Task* and *function*, *work* and *product*, *a work* and *to work*—these significations blend into a one-dimensional whole in the age of productivism. The ensemble of ends and means finds its temporal analogue in the coupling of *already* and *not yet* that denotes the degree to which available energy has been "actualized" in production. Ideally, a transitory discord between these markers of time would be resolved, and the result would be concretized immediately, that is, without delay and without undue mediations. The function *is* the product; the task *is* the accomplished work. That is another manner of time's disappearance in energy, not absorbed into happiness but frittered in the unfathomably fast rotations of potential potentiality and potential actuality, as requisitioned by the logistics of capitalist virtualization.

In the wake of the Aristotelian unmoved mover or of the Hegelian Spirit, production regulates the commerce of energy with itself. Both as subject and as substance, verb and noun: producer, product, to produce, produced. Hiding under the sheepskin of pragmatic concerns, the latest formalization of energy is a wolf, *more intensely metaphysical* than its forerunners, given the effectiveness with which it devours time, difference, and possibilities that disagree with its flat vision of the world.

Today's extreme and highly destructive energy dream presents itself as the sole possible reality (better: as the possibility of producing the one necessary and sufficient reality) that excludes other alternatives as unrealistic, utopian, unproductive. It claims for itself both the actual and the potential, censoring our reveries, fantasies, and previously unforeseeable imaginings or forms of seeing. Produced in vast "dream factories" (besides Hollywood, these include the entire ideological machinery of productivism into which everyone and everything is plugged without exception), they are harnessed for the sake of an ever more efficient and sweeping extraction-destruction-production of the future, which has no future. Yes, energy still dreams, though within strict repressive limits and under harsh conditions. The poverty of its dreams depends on the prohibition—its self-prohibition, more precisely—to maintain in being anything that is not and will never be productively at work. This for-nothing "is," for it, just that: nothing, nonbeing. Ontology is all the poorer when, having mobilized, actualized, or prepared for actualization everything that is and could ever be, it excludes unproductive possibilities from its ambit. Dreams, above all.

It is easy to surmise why *energy production* is so vital to the ideological constellation we are a part of. Not so much because, should it decline, the rest of the world economy would grind to a halt but because it is symbolic of how we live, be it as individuals, corporations, or states. Energy production is the production of actuality in our actuality, the activation of actuality, a hyperproduction, folded onto itself and consequently magnifying the effects of the prevailing destructive-extractive zeitgeist. This fold sweeps both reality and the dream into its midst and dictates the rules of the game: "it's my way or the (unusable) highway," the way of activation on the terms of production and the production of production or the highway of total deactivation, rest as death, the peace of the cemeteries.

Heidegger should receive credit for drawing our attention to the ontological overreach of production in ancient and modern forms of metaphysics alike. The "determinations adduced for *Sachheit* [thingness, reality]," he holds, and "*essentia—forma, natura, quod quid erat esse, definitio*—are obtained with regards to the producing of something."

From Plato and Aristotle through medieval Scholasticism to Kant, "production stands in the guiding horizon of this interpretation of whatness."[24] The *who* (Dasein) does not willfully foist productive parameters on the *what*; the *who* is, rather, under the sway of the *what*: "The apparently objective interpretation of being as *actualitas* also at bottom refers back to the subject . . . in the sense of a relation to our Dasein as an acting Dasein or, to speak more precisely, as a creative, *productive* Dasein."[25] As in the case of the artwork, we are privy here to a flattening of the categorial-existential divide. But, instead of endowing things with the characteristics of the *who*, as art does, the productive comportment defines who we are through the thingly *what*. Assuming that to act, to be an agent, to behave energetically is to produce, the actualizing action and, with it, the actor are contingent upon the demands of (and for) the intended product. And, contrary to what we might expect, the Greeks do not escape from Heidegger's accusing finger in this regard. The Platonic *eidos*, for instance, is the "anticipated look of the thing," "the anticipated look of what is to be produced by shaping, forming."[26] Themselves unproduced, the Ideas undergird the productivist ontology of the world here below. Their energy spins the world as the panoply of unsuspecting (dreamlike and sleepwalking) replicas of eidetic realities.

The metaphysical intensity of production that determines every *who* as a *what* surges in comparison to the unmoved mover or Spirit, where *who* and *what* still coexisted with more or less tension to be sublated and resolved, reminiscent of the fissures at the heart of ergon itself (*ergo*, energy). Medieval thought credited God with the identity of *existentia* and *essentia* in the plenitude of creative energy, as much thought as action, which will be one of my focal points in the next chapter. Lest I be misunderstood, I am not positing a return to these milestones of metaphysics as a sine qua non for resisting the productivism that assigns to us the function of producers (the function, which, to repeat, includes consumption) "objectively determined" by the correlative product. Equally futile will be the negation of productivism on its home turf, by dropping out of the system and becoming unproductive, an *Aussteiger*, as the Germans refer to it. The point is not to aspire to a higher-than-productive subjectivity, sometimes glorified in metaphysics, nor to be fully compliant with the political-economic program for which we are tiny fragments of an algorithm.

Writing that we have to work at the sense of *work*, I propose to unmoor it from production and to work through (in the psychoanalytic sense) our compulsion to fabricate the world and ourselves, without sparing the "progressive" performative, discursive, narrative, and other methods of generating *what is*. With the environmental crisis upon us, it is necessary to dream up another energy, another enworkment where humility and taking charge, accepting the given and elaborating it, belong together. How to put this dream to work outside the exigencies of producing and reproducing a reality that, day by day, appears to be more and more virtual? The chapters on theology, economy, psychology, politics, and science that follow contain hints of an answer. Yet it would be also wrongheaded to predicate the thinking of energy on a search for solutions, seeing that a question resolved and pacified in the response willy-nilly capitulates before the machinery of production, with its "positive" outcomes. What we can do (and this modal verb *can*, promising potency, as much as the active *do*, should not be taken for granted) is let another energy work and dream, as it gushes forth from the fault lines of the productivist worldview.

2

THEOLOGICAL MUSINGS

From Aristotle we have learned that the energy of the plenum, which is also the plenum of energy where nothing is lacking, is the actuality of powerlessness, energeia rid of the still unfulfilled capacities of dunamis. On the one hand, resorting to the qualifier *powerlessness*, language has proven somewhat inadequate to the mission of giving a voice to the charge of this energy, brimming with positivity. On the other hand, the rejection of power is consistent with the method of negative theology that deems God to be inaccessible to finite human beings, save for an oblique *via negativa*. In my interpretation of *Metaphysics* 1048a, 31–32, I have alluded to the feasibility of receiving in this way Aristotle's brittle definition of energeia as "not what we mean by dunamis [μὴ οὕτως ὥσπερ λέγομεν δυνάμει]." More often than not, however, monotheistic theologies have been mesmerized by divine omnipotence that precludes the power to be powerless. They have conveniently cast aside the old argument that omnipotence gathers into itself all the potency, every *potentia*, and therefore waxes impotent in the face of energetic actuality. An all-powerful God is powerless to embrace the powerlessness of *energeia*. As we shall soon see, the Scholastic (notably, Thomist) translation of Aristotle's word into *actus purus*, denoting perfection, has done much to cement this self-contradiction.

The proliferation of notions such as "God without God," "God beyond God," "the weakness of God," or "a trace of God" in late twentieth- and early twenty-first-century philosophy and theology has been reversing the dogma of omnipotence. By challenging the attribute of divine sovereignty, His supreme and unreachable authority, Emmanuel Levinas, Jacques Derrida, Gianni Vattimo, John Caputo, and others have begun to dream of a departure from the monarchical God handed down to us from Scholasticism. Within the context of Christian theology, this move is understandable in (the) light of the incarnation that is, in the words of Vattimo, "God's abasement to the level of humanity, what the New Testament calls God's *kenosis*," which "will be interpreted as the sign that the non-violent and non-absolute God of the post-metaphysical epoch has as its distinctive trait the very vocation for weakening."[1] But is weakening a quantitative fluctuation in or a qualitative about-face of divine energy? Is it still ensconced in the divine economy of ends and means, where the death of God is the means potentiating the end of salvation? How is this diminution activated? What or who energizes it? Is nonviolence a mask of quasi-nihilistic passivity (e.g., submission to one's fate on the cross) or is it an upshot of another energy, no longer reliant on the routines of extraction-destruction?

Trouble is that, for all their commendable and sensitive features, theologies of weakness sweep the issue of energy under the theoretical rug. They join Heidegger in a silent consensus on the proper place of energy among the keywords of metaphysical ontotheology that, collectively, make up a catalogue of violent absolutes. The obduracy of this conviction has to do with the tacit preponderance of power-dunamis over the powerlessness of energeia in the theologies of weakness themselves. Weakness portends exhaustion, rather than plenitude, the abatement of power and of metaphysical violence in a weary God and an ever more fatigued humanity, tired, in the last instance, of themselves. But what if kenosis and Christ himself were construed as the exhibitions of God's energy, of positive powerlessness that has no place on the metaphysical scales, of the divine becoming the working principle and the work? Both higher and lower than "the highest poverty" of monasticism and asceticism,[2] of abasement and abdication, kenosis is energeia, unrecognizable insofar as it points away from the absolute, from violence, or from mere production. The very Trinitarian "structure," with its inner division (economy) and *stasis* (rest *and* strife), would then be a sign of this energetic overflow.

Although his writings inspired the theologians of weakness, Friedrich Nietzsche suspected those who affirmed and enjoyed powerlessness of a masquerade, where the will-to-power was buttressed through circuitous and reactive means. That was Christianity's "stroke of genius," *Geniestreich*: "God sacrificing himself for the guilt of man, God himself exacting payment of himself, God as the only one who can redeem from man what has become irredeemable for man himself—the creditor sacrificing himself for his debtor out of *love* (is that credible?—), out of love for his debtor!"[3] The point is well taken, inasmuch as it locates the Christian "stroke of genius" within a totalizing closure, which incorporates the elusive remainder of powerlessness in the power of God. But it falters, provided that the contrasts power/powerlessness and activity/passivity stay at the level of unexamined presuppositions, amnesic of the ancient energy—not of restlessness but of rest.

Active human beings "simply *felt* themselves to be 'happy' . . . ; and as full human beings, overloaded with power [*als volle, mit Kraft überladene*] and therefore *necessarily* active, they likewise did not know how to separate activity out from happiness," while in the weak, "the powerless, oppressed [*der Ohnmächtigen, Gedrückten*]," happiness "essentially appears as narcotic, anesthetic, calm, peace, 'Sabbath,' relaxation of mind and stretching of limbs, in short, *passively*."[4] Nietzsche is, no doubt, intimately acquainted with the Aristotelian luggage of *happiness* (*eudaimonia*) tied to *energeia* in the fullness of its accomplishment. But, if the power-laden humans he admires are its legatees, then they must experience, simultaneously, "active" self-affirmation and "passive" rest, calm, peace. These adjectives may indeed be applied in the same measure to energeia's rest and brimming, effervescent action. Nietzsche is modern, all-too-modern to acknowledge that other energy, thanks to which the separations between the weak and the strong crumble. For him, fullness means being "overloaded with power," an excess of dunamis, of potentiality, not the actuality of energeia. Given his conceptual tools, he is unable to welcome powerlessness (in particular, that of God) and is obliged to put it on display as a deficiency that, perversely, endeavors to rule the world. In the standoff between the weak and the strong, there is only activity and its slowing down or cessation, which aims to infect and subdue the effervescent self-affirmation of those who are happy. The activity *of* powerlessness is explained as a reaction to the independent exercise of power, via Nietzsche's unspoken psychic modification of Newtonian physics, with its third law of motion: for every action, there is an equal and opposite reaction.

All this is but the tip of the iceberg that is the metaphysics of the will-to-power. Escalating the virtuality of dunamis, this master-concept of Nietzsche's philosophy doubles potentiality into "will" and "power," a doubling that culminates in potentiality's relation (striving) to itself, the will willing its own increase, a power to have power. To the extent that actuality and what is in it persist, they ensue from this self-relation, which is the inverse of the unmoved mover's self-intending energy Losev teased out from Aristotle. A double potency and a double forgetting of itself, energy dreams of its increase actualized in the growth of power, that is, never quite actualized. Weakness is a slackening of potentiality's striving to itself, a blockage in the *to* of the will-to-power, and a perverted continuation of this movement in the name of its obverse (the negation of power, powerlessness). Thus kenosis will have been the most cunning and sublime instance of this perversion, an unfathomable increase in the Christian will-to-power under the pretext of its abnegation.

The kind of divinity Nietzsche favors cannot be found within the confines of the Judeo-Christian heritage. Among the fragments that comprise *The Will to Power*, we come across a manifestly Aristotelian concession to the meaning of god. "The sole way," Nietzsche notes, "of maintaining a meaning for the concept 'God' would be: God *not* as the driving force, but God as a *maximal state*, as an *epoch* [*Gott* nicht *als treibende Kraft, sondern Gott als* Maximal-Zustand, *als eine* Epoche]—a point in the evolution of the will to power by means of which further evolution just as much as previous evolution up to him could be explained."[5] In other words, a point of completion within history, a total desaturation of dunamis in the fullness of energeia, jettisoning the claims to power, force, drive. Even a deistic God, who gives the first creative push to the world from which he then retracts, would be too "active" on this view, protesting against the conceptual translation of *prima causa* into theological categories. God is not what (or who) happens atemporally *before* history but *after* it, provided that the "maximal state" of closure is grasped *within* historical becoming, as "a point in the evolution of the will to power." Be this as it may, when god is thought of as a state or an epoch, and not as a driving force, the contrast between passivity and activity, fundamental to Nietzsche's philosophy, starts fading away.

Tellingly, Nietzsche goes on in the same Fragment 639 of *The Will to Power* to discuss energy once again through an Aristotelian lens (reinforced by "mechanistic" and "economic" factors) in terms of the concert of stability and change: "Regarded mechanistically, the energy of the totality of becoming remains constant; regarded economically, it rises to a high point and sinks down again in an eternal circle [*Mechanistisch betrachtet, bleibt die Energie des Gesammt-Werdens constant; ökonomisch betrachtet, steigt sie bis zu einem Höhepunkt und sinkt von ihm wieder herab in einem ewigen Kreislauf*]."[6] But if energeia congruous with the will-to-power is the overarching difference between mechanistic and economic attitudes, then it cannot help but enter the chains of cause and effect (mechanism) or means and ends (economy). Powerlessness is reduced to a moment in energy's economy, its "sinking down" in anticipation of a future upsurge. That is Nietzsche's blind spot: the will-to-power occludes the fact that the "energy of the totality of becoming," *die Energie des Gesammt-Werdens*, whose name or misnomer is "God," is totally powerless, *adunamon*, prior to the conceptualizations of weakness and strength.

Reading the philosophers and theologians of weakness against the grain, I have been trying to align their thought with the powerlessness of the Aristotelian energeia. Still, the resistance to energy we alight upon in postmetaphysical circles is exceptionally difficult to overcome. What it balks at, above all, is a far-reaching theological distortion of Aristotle, subsequently reprojected onto ancient Greek thought. I am referring, of course, to Thomas Aquinas and his infelicitous translation of energeia into actus purus, a pure act: "Something is knowable in so far as it is actual [*in actu*]. But God is a pure act without any admixture of potentialities. Insofar as he is in himself, he is supremely knowable. [*Deus, qui est actus purus absque omni permixtione potentiae, quantum in se est, maxime cognoscibilis est.*] Yet what is in itself supremely knowable is not within the scope of intellectual cognition, exceeding by far the power of an intellect" (*Summa Theologiae*, I.12, 1). The purity of "pure act" is its freedom from the admixture of potentialities. The Thomist God is energeia minus dunamis, a pure energy working in excess of our finite embodied understanding, yet simple and maximally knowable in itself. He is an activity that knows no equal and no opposite.

God's perfect and accomplished ability, the actus purus that he is, results in severe limitations that actually undermine his omnipotence, frustrating St. Thomas's earnest exertions. For instance, "it is impossible that matter would be in God. First, because matter is potentiality [*in potentia*]. On the other hand, God, who is a pure act has nothing of potentiality [*quod Deus est purus actus, non habens aliquid de potentialitate*]. It is, therefore, impossible that God would be a composite of matter and form. . . . He is by his essence form [*Est igitur per essentiam suam forma*]" (*Summa Theologiae* I.3, 2). The essential formalism of God posits him over and against the world, excluding the imperfections and incompletions of matter from his essence. The work of actus purus will, therefore, be the work of acting upon matter by forming it, by impressing the power and sovereignty of pure form on it, albeit without taking part in it.

As energeia turns into *actualitas*, much of the Greek word's intriguing opacity is wiped out. First, the *en-* of *enworkment* is erased, leaving behind a variation on work, the substantive of the verb *agere*, "to act," bereft of spatial and temporal coordinates. Second, actus purus is a pure noun that, despite its connection to the verb *agere*, specifies the essence, or the substantive being, of God. It excludes the connotations of a process, of activation, of an undertaking. Third, the noun is indexed to another noun, namely to the supreme Actor-God. Bearing the imprints of his activity, the actuality of the world is passively receptive to actus purus. Contrary to energeia that has swept into its midst the ongoing work and the works (the working and the worked upon), the God of

St. Thomas lays the foundations for a riven world of active subjects, on the one hand, and passive objects, on the other. Henceforth, the things themselves, the works, will not be permitted to work but will have to tarry until the worker's forming activity dispenses to them their actuality, imitating God's formal and forming treatment of matter.

The other significant disjunction between energeia and actualitas is that only the latter makes inroads into causal relations. Set to work or trapped in a work, energy is always in the middle, *in medias res*, in the midst of beings without beginning or end. The Aristotelian *first actuality*, too, is not chronologically prior to the world (or to a body), for it arises from their self-relation. Not so in the case of actus purus, which is, by the same token, an *actus primus*, the first act, the active cause of finite existence. God-the-Actor is a *causa sui*, the self-caused Cause of creation, the first energy energized by itself alone. By forming the potentiality of matter, the subject causes things (essentially, formed matter) to come into being. "Even if one adds to causality, which only poses an effect in being, a self-positing of being . . . , *actualitas* is still, as Heidegger says, thought '*im Hinblick auf die* causalitas,' from the perspective of causality."[7] Seen causally, the *who* and the *what* obey the laws of production: the self-production of God, the prototype of "autonomous" subjectivity, on the one hand, and the production of things, on the other. Creation is reduced to the materialization of divine formal energy, sanctioned by the creed of productivism.

———— ❧ ————

The world is a laboratory of creation, fabricated with its sundry contents by its Maker. This image recurs in disparate theological currents and epochs of Christianity, from Jakob Böhme's 1622 credenda laid out in *De signatura rerum* to William Paley's "watch analogy" that shores up the argument from intelligent design in his 1802 *Natural Theology*. Creation is seen as a pale trace of productively expended divine energy, an Ariadne's thread that, if followed correctly, could lead us back to the Maker.

Anteceding Böhme and Paley, St. Augustine had devised a system of creation as fabrication in his *Tractates on the Gospel of John*. A carpenter, he states in an evidently Platonist mood, first makes a chest in his "creative knowledge." There, "it exists invisibly; in the product it will exist visibly." The Wisdom of God (*Sapientia Dei*) closes the gap between the intention and its realization. Divine knowing and doing (or, in our terms, theory and practice) are an organic unity. The production of actuality *in* God is prior to the creative act itself. The *en-* of divine energy, indifferent to spatial and temporal interiorities, is recovered at the level of God's sublime inwardness: "the Wisdom of God, through which all things were made, contains all things in accordance with his creative knowledge before he constructs all things [*Sapientia Dei, per quam facta sunt omnia, secundum artem continet omnia, antequam fabricet omnia*]" (*In Evangelium Ioannis* I, 17).

Observe how Augustine designates creation: reality is comprised of *facta* (from the Latin *facere, to do* or *to make*), works made by God that live in him before the actual fabrication of the world. Our objective-empirical "facts," the idols of the techno-scientific outlook, are, in a convoluted way, the unacknowledged vestiges of objectified divine energy. To be fair, Augustine is careful to distinguish the fabrication or construction (*faber, fabricet, fabricari*) of the world that bears resemblance to production from the life (*vita*) of things in God: "whatever things are made through this creative knowledge are not immediately life; but whatever has been made, or whatever is a fact, is life in him [*sed quidquid factum est, vita in illo est*]. . . . You see the sun and the moon; these, too, exist in his creative knowledge. But externally they are bodies; in his creative knowledge they are life [*ed foris corpora sunt, in arte vita sunt*]" (*In Evangelium Ioannis* I, 17). That the worldly "facts" guide us back to spiritual realities is the ground rule of Augustine's allegorical hermeneutics. What interests me, though, is something else, namely the contrast between the life of things in God and their objective existence alongside other artifacts of creation. Accounting for the external and the internal, the body and the life of things, Augustine pieces together the two halves of ergon: the verbal sense of work and the released works-products. But he does so by transferring the ambiguity of energy onto symbolic grounds, where the bodies of works-products are signposts for the superior spiritual reality of a living work.

Only in God is there energy that is alive, and the bodies that do not hark back to it are as good as dead. To reiterate, the *en-* of energeia names, from the Augustinian vantage point, divine interiority, the inhood of God, superseding that of time and space, since *vita in illo est*. Life before creation is not, *à la* Hegel, an emptily abstract universal but the always already actualized vigor of and in God. The exteriority of facts, of created bodies, those that "externally are bodies" (*foris corpora sunt*) are the works without enworkment, drained of energy and prone to decay as soon as the umbilical cord connecting them to God is cut. Therefore, the productivist paradigm of creation goes no further than the external appearances of dead works (the existing facts) while shunning the energy of *Sapientia Dei*, which maintains the life of things in advance of the world's "fabrication." If nowadays productivism is rampant, that is because, having proclaimed the death of God and unfastened the materiality of bodies from their spiritual sustenance, secular humanity has not bothered to ferret meaning out of corporeality itself. Our total mobilization of energy is geared toward the production of facts, of the a priori dead bodies of things, commodities, plants, animals, and humans, infinitely malleable, manipulable, and inconsequential.

Rather than doctrinaire hairsplitting, irrelevant to the secular world, the study of theology may explain the axiological and epistemological patterns that prevail in actuality. (*Secular* and *world* are, themselves, theological constructs.) Besides the concept of facts that postmetaphysical discourses borrow unawares from Christian theology, we cherish the dream of partaking in God's creative knowledge as we shrink the time interval between the storage and the release of energy. Firmly yet also unconsciously entrenched in a productivist take on creation, modernity wants human knowledge instantaneously to yield fruit in a skewed *imitatio Dei*. But contrary to divine actus purus that does not suffer the division between theory and praxis, for us, who yearn to bridge them, they remain worlds apart. In the global climate of technocracy dominated by "economic rationality," the false reconciliation of the two confirms, at bottom, the tyranny of action over thought: knowledge is to be immediately productive according to pragmatic goals that, properly understood, are means without an end. The type of thinking the system consents to set to work, to energize economically (i.e., by funding research proposals consonant with it), is one that will produce better research statistics, more cost-efficient construction materials, chemical fertilizers, and so forth. That is not creative knowledge but, at best, a knowing creation—procreation, reproduction, perpetuation—of the destructive status quo.

The more "energy resources" are secured, the faster they are spent, burned together with the rest of the world in the bottomless pit of productivism. Our staunchly held prejudices concerning matter as dumb and inert prompt us to set it to work by negating it and, through this negation, draw energy from it at an ever accelerating pace. Not even plants are spared the drive to minimize the time between the storage and discharge of energy. Fruit are plucked unripe from trees, gassed into ripeness, or placed into special containers designed to continue ripening on a supermarket shelf or at the consumer's home. Their *erga* (both verb and noun), their energy, their enworkment are put to work so as to mollify agrobusiness's deranged dream of decreasing production time and increasing profit margins.

Offering an alternative, nonproductivist account of creation, Meister Eckhart carves another approach toward divine energy. In Eckhart's mysticism, God is an eternal fire, boiling within himself (*inneres Kochen, bullitio*), while creation is the spillover of his fiery plenitude into what will have become the world, a boiling out or a boiling over (*Überkochen, ebullitio*) of divine flames. "God as good is the principle of the 'boiling over' [*ebullitio*] on the outside; as personal notion he is the principle of 'boiling within himself,' [*bullitio*] which is the cause and exemplar of the 'boiling over.' Thus, the emanation of the Persons in the Godhead, the cause and exemplar of creation, is prior."[8] The unthinkable fullness of his energy, God's "boiling within himself," revives the memory of energeia. Infinitely fertile and exuberant, it cannot be detained with its confines (it has none!), which is why it engenders the world freely, non-productively, in an act of grace. At the same time, the energetic excess of God over himself that creates the actuality of existence is a repetition of an earlier *ebullitio*, whereby God engendered more than himself in himself. The Trinity is a rehearsal for the creation of the world, "the cause and exemplar of creation," as Eckhart writes. Christ-the-Son is the energetic effect of God-the-Father, if only an effect inseparable from the cause. In keeping with pyrological imagery, the Father and the Son are "fire and heat," and therefore "extensions of one another."[9]

Eckhart's Trinitarian vision thus warrants a conception of divine energy beyond the metaphysics of pure and eternal activity that governs Thomist thought. The God of ebullitio is the energy of undisturbed rest and of creative activity, remaining in itself while overflowing itself into the other Persons of the Trinity and, subsequently, into the world. More than himself while one with himself, God is both God (*Gott*) and Godhead (*Gottheit*): "God acts, while Godhead does not act. There is nothing for it to do, for there is no action in it. It has never sought to do anything. The difference between God and Godhead is that one acts and the other does not."[10] A sensible exegesis of these lines would contend that Eckhart is differentiating between divine essence, which is in perpetual repose, and the active-creative divine energy, prone to movement. The differentiation between essence and energy (the latter synonymous with existence) was undeniably central to Byzantine theology and, in its footsteps, to Eastern Orthodox Christianity. And yet it would be ill-advised to overlay this rushed hermeneutical decision on Eckhart's text. Energy need not be synonymous with activity, as Aristotle has argued, but it is just this synonymy that we have asserted in our "sensible" exegesis. The statement that, as far as Godhead is concerned, "there is nothing to do" does not presuppose that the essence is idle or inert. On the contrary, Godhead concentrates the Aristotelian energeia beyond the active disposition of God.

The same confusion besets the concept of divine excess, or ebullitio. Consider ancient Greek thought on the subject of excess. The Platonic Good exceeds the rest of the Ideas and is, therefore, beyond both knowing and being, ἐπέκεινα τῆς οὐσίας (*Rep.*, 509b). It is a good excess, or, indeed, excess as the Good. Aristotle, vice versa, pleads for moderation in *Nicomachean Ethics*, locating virtue in the middle between two excesses, the extremes of deficiency and surfeit. Are the teacher and his student contradicting each other? Or are they talking, as I think they are, about two aspects of excess (let us label them *excess in itself* and *excess for us*), mindful of human limitations when handling such dangerous, untamable things? The energetic excess, energy qua excess, similarly plays or works itself out on two parallel levels. In itself, it is sufficient; for us, it is never satisfactory.

Plato's Good, Aristotle's unmoved mover, Eckhart's God or Godhead, and Hegel's Spirit are the archetypes of energetic excess in itself. (With Augustinian Platonism for an intermediary, Eckhart superimposes the good onto God—"God as good is the principle of the 'boiling over' [ebullitio]"—which is to say that there is something of the Platonic Sun, itself analogous to the good, in the burning of divine fire.) After the screens of ontotheology and Western metaphysics have come down, it has fallen to us to translate the self-subsistent fullness, rest, and positive powerlessness of past thought into an excess for us, saying about it, perhaps for the first time, *Sufficit!* Behind these screens, we do not glimpse a nonmetaphysical image of reality, for even *facts* are the unwitting mementos of creation. There we finally manage to discern the ongoing work of finitude (limiting and ending), the work that has been the furtive energy supply of metaphysics since the inception of philosophy. *Energy Dreams* matures (or mature) according to this script: my wager is that the Aristotelian energeia is a decisive link in a long chain of previously exalted and subsequently denigrated notions awaiting a fresh interpretation after the closure of metaphysics *and* its deconstruction.

Little known in the West, Gregory Palamas, a fourteenth-century Byzantine monk who would rise in Church ranks to become the archbishop of Thessaloniki, exerted an enormous influence on Eastern Orthodox theology and philosophy. (In the twentieth century alone, leading Russian philosophers Vladimir Lossky, Vladimir Bibikhin, and Sergey Horujy have responded, in one way or another, to Palamas's teachings.) Among his notable achievements is defending a rigorous division between (and the economy of) the essence (*ousia*) and the energies (*energeiai*) of God, particularly in *Hyper tōn hierōs hesychazontōn, For the Defense of Those Who Practice Sacred Quietude*, more widely known as *Triads* after the structure of the text written and arranged in three series of three treatises.

Insistence on the contrast between divine essence and energies was not original to Palamas, nor to Eckhart for that matter. The Capadoccian Fathers of the fourth century, whom the Byzantine monk extensively cites, had envisioned it before him, as did Cyril of Alexandria, Pseudo-Dionysius the Areopagite, and Maximus the Confessor. Palamas's achievement was to sharpen the argument in a bitter debate with Barlaam of Seminara, an Aristotelian Scholastic who accused his opponent of heresy, and to ally it to the spiritual practice of Hesychasm (from the Greek *hesychia*: "stillness," or "quietude"). The six patriarchal councils that took place in Constantinople between June 1341 and May 1351 ruled in favor of Palamas, cementing the schism between Roman Catholicism and Greek Orthodoxy.

Even as the Scholastics were defending the Thomist creed that God's essence and existence were one and the same, Palamas underscored the dissimilitude of divine essence and energies. Structurally, Palamist energies occupy the place of existence, an idea that *eo ipso* challenges the Scholastic dogma. They are the works (erga) of God, not as the objectifications of his will in creation but as his active qualities: prescience, providence, self-contemplation (*Triads* III.ii.6). The energies are made of the erga that capacitate (for instance, divine foreknowledge) and that put the essence to work. And yet the activity of God's energies rebels against the requirements of ontotheological metaphysics. In this variance lies the promise of Palamist theology.

Worlds apart from the simplicity, hiddenness, eternity, and self-absorption of divine essence, the energies are multiple, manifest, partly finite, and turned toward exteriority. They are "revelatory" (III.ii.7); every one of them "entirely manifests" divine essence, "His essence being indivisible" (III.ii.8); in the form of the eternal glory of God, they are participable "for that which in God is visible in some way is also participable" (III.ii.13). Shaping up in Palamas's postulates is a phenomenology of divine energies that is on par with the nonphenomenality of his essence. The metaphysical procedure for tackling phenomena mandates that the energies with their multiplicity, exposure, sharedness, etc. be conceived as a lesser mode of the perfectly unitary, withdrawn, and self-absorbed divine essence. Palamas is, nonetheless, adamant that the essence of God is both irreducible to the energies and wholly evinced in each of them. I would go so far as to say that the correspondence of energies to existence anticipates Heidegger's ontology: the *ousia*-essence-being of God is not a supreme being, but the being *of* his *energeiai*-existences-beings. With Palamas's interference, something incredible ensues in the universe of metaphysics: existence activates essence.

More impressive yet is the submersion of divine energies in finitude. Uncreated, some can have both a beginning and an end, while others (e.g., the prescience of God) can have an end but not a beginning. The baffling idea of uncreated works both challenges the productivist take on creation and opens unto secular thought—something that earned Palamas the charges of heresy. "Indeed," he writes, "beginning and end must be ascribed, if not to the creative power itself, then at least to its activity [πράξεως], that is to say, to its energy directed towards created things [τα δεδημιουργημένα ἐνεργείας]" (III.ii.8). The exercise of creation is as finite as creation itself. Creative power, referring to one of the divine energies, descends to the level of temporal existence, which will emanate from it. Put in another way, the energy of energy, its very essence, the activity or activation of energeia, is configured existentially. As is the very notion of *essence* that, no later than it is named, is transformed into vanishing energy, encrypting and giving itself to us, giving-encrypting itself in a word with which it does and does not coincide: "even this name 'essence' designates one of the powers of God" (III.ii.11).

Palamas, to be sure, preserves the surface/depth divide in the sundering of God into divine essence and existence. Yet, he also thwarts it by situating energy on the phenomenal surface and by zeroing in on something like *the surface of depth*, the energetic token of essence. His *energeiai* are not containable in themselves; they are not withdrawn (withdrawal is the prerogative of essence) but sharable and exposed, turning ever inside out, handing themselves over to the other. These energies on and of the surface need not be violently extracted, seeing that they generously and graciously irradiate toward the exterior. Being energized by them is only a matter of finding the right method for their adequate reception.

Palamas presents Hesychasm, the series of spiritual practices he upholds, as a way to receive the energies of the divine surface. A hesychast cares for the bodily interiority of the heart, the organ to be quieted down through meditation and the control of breathing. There, in a calm heart, divine energies will set themselves to work. The outward manifestations of God touch humans in the hidden niches of the body that are exterior compared to the thoroughgoing retraction of divine essence. Before the divine we always wear our hearts on our sleeves, *a flor de piel*, as they say in Spanish: "like a flower of the skin," or, less literally, "skin-deep." Criticizing Hellenism for wishing "to make the mind 'go out,' not only from fleshy thoughts, but out of the body itself, with the aim of contemplating intelligible vision," Palamas's counsel is that one "recollect the mind not only within the body and heart [του σώματος καὶ τες καρδίας], but also within itself" (I.ii.4). To make this happen, it is imperative to avoid distractions and "to control the mind together with the breath," such that "the to-and-fro movement of the breath becomes quieted down during intensive reflection, especially with those who maintain inner quiet in body and soul [των ἠσυχαζόντων σώματι καὶ διανοία]" (I.ii.7). Although one might misjudge the quietude of Hesychasm as inactivity, the inner concentration of the mind, of the heart, and of the breath aspires toward a down-to-earth instantiation of the Aristotelian energeia, harboring plenitude and rest. If hesychasts "practice a spiritual Sabbath, and, as far as is possible, desist from all household work [ἀπὸ πάντων των οἰκείων ἔργων καταπαύοντες]" (I.ii.7), they do not indulge in lazy navel-gazing, as Barlaam of Seminara frequently joked, but calm down dunamis with its relentless becomings in the midst of corporeal existence.

As he defends and engages in hesychastic practices, Palamas dreams of divine energies within himself, in his bodily and mindful core where they would set themselves to work. Discovering and welcoming them within himself, the hesychast is moved . . . to be still. In Palamas, Aristotle's energy dream recurs and leads a Christian afterlife, proving to be an irrepressible aspiration.

Having made a brief detour through the East, we should linger with St. Augustine a little while longer. His writings and sermons mark a crucial period in the history of the West (and, by extension, of metaphysics) when works became allegorical in the *subtilization of energy*. Defying the literalness of Judaic law, Augustine stressed the law *naturaliter in corde conscripta*, "written naturally in the heart" (*Ep.* 107, 15), not foisted from the outside. His, however, was not the heart of the hesychasts, the actual bodily organ trained to receive stillness through a conscious practice of breathing and to recollect the practitioners within themselves. The Augustinian heart is a psychic interiority, loosely connected to the body with allegorical ties and contributing to the subtilization, or sublimation, of energy channeled into spiritual outlets. It is, in truth, the same heart that St. Paul appeals to as he rejects the Hebraic circumcision of the flesh and substitutes for it that of the heart: of thoughts, wishes, intentions, and desires to be tamed and repressed.

Pauline interiority is the site where faith is energized by love. "Faith sets itself to work only through love [πίστις δι' ἀγάπης ἐνεργουμένη]," he writes in *Galatians* 5:6. The enworkment of faith operates with an invisible "product," matching the place of the spiritual heart subtracted from physical space. Away from corporeal sight, love (*agapé*) binds the believer to the community of believers, to the Church, and to the body of Christ. Faith (*pistis*) imbibes the energy of love, thanks to which it can metamorphose into a religion, a re-binding of those who share it.

Painted in broad brushstrokes, the scene is set for the subtilization of energy in Augustine, who resolutely argues that the works, *opera*, are not in the first instance physical creations but acts of spirit, and, hence, of faith activated by love. In *Confessions*, he instructs that "fruits of the earth are to be allegorically interpreted as meaning works of mercy [*dicebamus enim eis terrae fructibus significari et in allegoria figurari opera misericordiae*], which are offered for the necessities of life from the fruit-bearing earth" (XIII.xxv.38). Explicating the biblical injunction "Be fruitful and multiply!", Augustine transcribes "fruits of the earth" into "works of mercy" (*opera misericordiae*) and so replaces physical procreation with the multiplication of spiritual works. In the heat of his polemics, he appropriates the energy of the fruits themselves, along with that of the literal surface of the text, for the sake of spirit. Vegetal nature that sets itself to work in the plant and actualizes itself in the fruit is spiritually domesticated, converted into the subtle energy of love, faith, and hope. In its becoming-virtual, energy is uprooted from the terrestrial support: *terrae fructibus*, the fruits of the earth, are dispossessed of their literal significance precisely inasmuch as they belong to the earth, devalued in comparison to the ethereal spiritual realm. They grow meaningless, their energy devoted to growth waxes meaningless, so that works of mercy would be endowed with indisputable significance.

The Augustinian energy dream feeds on the very material, finite, embodied energy it blocks, debases, and forbids, seeing that, without the literal fruit, there is no allegory of the works. Devalued and instrumentalized, vegetal, animal, and human bodies (physical powers, capacities, enactments, or works) are subservient to the demands of the soul to which they are sacrificed. Given this rift, energy is divided against itself, separated from its corporeal substratum and denied the chance for a direct expression there. When used well and correctly, in *bonus et rectus usus*, bodily energy will be donated in sacrifice to spirit: "The body, then, which, because it is inferior the soul uses as a servant or instrument, is a sacrifice when it is used rightly and with reference to God. And, if this is so, how much more does the soul itself become a sacrifice when it directs itself to God so that, inflamed with the fire of His love [*ut igne amoris eius*], it may receive His beauty and be pleasing to Him, losing the form of worldly desire and being reformed immutably by its submission to Him!" (*The City of God*, bk. 10, ch. 6). The key to availing oneself of one's body well is in immolating it on the altar of the soul; the correct use of the soul is surrendering it to God. But that is not enough. As the Augustinian believers funnel the energies of their bodies and souls to God, they realize that the energy of God's love has been moving them on the right track all along. It carries on the enworkment of faith that we have espied in the Pauline text and is featured here as the perfect example of infinitely renewable energy, a fuel that cannot be depleted, a combustible that drives the soul *igne amoris eius*, "inflamed with the fire of His love." Not only do finite bodies and souls undergo devaluation and debasement, but they are also revealed as derivative and inessential relative to the inextinguishable divine bonfire into which they are thrown. Works of mercy *are* these "true sacrifices"—*vera sacrificia opera sint misericordiae* (*The City of God*, bk. 10, ch. 6)—of the material to the ideal, to the Energy of (or above or behind) every energy in the world. Material destitution nourishes the dream of energetic indestructibility.

The ascription of the highest, the first and the last, energy to God is in agreement with the Judeo-Christian thesis that all being is through and thanks to God. But allegory, Augustine's weapon of choice, says otherwise: it must take its cues from the literal sense that is to be over-stepped and decried as slovenly. Allegory is the other vis-à-vis the same that is the letter of the text. Its energy beams from the negation of this semantic surface, the text's body, which—its travails, torsions, and distortions notwithstanding—it cannot discard in toto. Whereas an allegory turns meaning against itself, converting extended (embod-ied) sense to spirituality, the subtle energy it releases, the purified energy cleansed of literalness, works by making sure that other materi-al-energetic configurations do not set themselves to work so long as it is active. A purely spiritual energy, which dreams of its detachment from the body compelled to shed "the form of worldly desire," is, psychoan-alytically speaking, repressive. Its erga are, consequently, impregnated with negativity and epitomized in the works of mercy that are "true sacrifices." The actuality of spiritual-allegorical energy entails, far from the positive plenitude of energeia, the self-abnegation of the body for the soul and of the soul for divinity. Submitting to it, human works and their enworkment are prompted to renounce themselves *for their own good*, so as to be preserved in God.

As though an enormous historical distance did not separate us from early Christian theology, Augustine's words ring in our ears like a familiar tune, their rationale scrupulously secularized and successfully digested into our economic and political realities. Energy divided against itself is the mainstay of civilization, with its basic or surplus repression of desire, as Freud and other psychoanalysts have attested. In twenty-first-century Europe, austerity politics, with its "enlightened" technocratic demeanor, relies on personal and collective sacrifices that are demanded of (South)

European citizens and that are virtually identical to the Augustinian good works. Elevated to the status of God, economy exacts physical and psychic energy from human and nonhuman beings, transforms that energy into one of spiritual variety, and gives us the impression that we cannot survive unaided by its workings and prescriptions of the *bonus et rectus usus* (read: an efficiently productive use) of our time. And this is not to mention that much economic activity or activation is predicated on "faith" (credit or trust) and inspired by the "love" of profits or of oneself (Adam Smith's self-interest). Between the literal and the allegorical, the energy of self-abnegation is supposed to impart spiritual vitality to the economy both in accord with Protestant ethics, whose consanguinity with capitalism preoccupied Max Weber in the early part of the twentieth century, and with the Augustinian construal of sacrificial works.

But even in Christianity energy does not have to undergo the crass Platonic partitioning it suffers at the hands of Augustine, for it can be shared in a *synergetic* relation.[11] Paul hints at a nondomineering rendition of divine love in 1 Corinthians 3:9, where the believers are the "co-workers," *sunergoi* of God: "For we are the co-workers of God [θεοῦ γάρ ἐσμεν συνεργοί]." Rather than passively energized by the divine Cause, we are at work together with God, partaking of the same energy, or committed to the same objective (say, the good), around which our energies aggregate with those of divinity. Pseudo-Dionysius the Areopagite regarded the energy of Christ as *theandric*, θεανδρική ενέργεια, an intersection between the energies of God and that of humanity. For Maximus the Confessor, *theandriké energeia* boils down to identity in difference, "for by the word 'theandric' the teacher [Pseudo-Dionysus] obviously refers periphrastically to the double energy of the double nature.... When he wants to designate monadically the union of natural energies, he says 'this energy,' which does not harm the natural difference between them, just as it is not by identity of essence that they form an inseparable union" (*Op.* 7, 85A). The *with-energy* or the *energy-with* of *syn-ergeia* is an unbound bond that, careful not to suffocate existence in the nets of essence ("it is not by the identity of essence that they form an inseparable union"), airs the ties of re-ligion. It allows for the independence of two interrelated energies without subjugating the one to the other.

Just as a human incarnation can freely and synergetically join the energy of God, the body and the soul can also cling to one another on the hither side of instrumental considerations, subservience, and the Augustinian hierarchy. Layers of meaning, after all, coexist in any text: literal and figurative, saying the same and the other thing, articulating surface and depth. Synergy does not authorize the sacrifices of the body for the soul and of the soul for God; it empowers working, or setting-to-work, for a common goal and toward the same end, or else reposing in the actuality of what has been brought forth.

Being-at-work together, or being-at-work-with, is *sharing* energy, if only with *that upon which* one works. Contrary to the finitude of physical resources and the eternity of the theometaphysical blaze, synergetic transactions neither deplete nor maintain unchanged the "quantum" of energy but revel in its excess, overflow, plenitude, movement and rest, movement in rest and rest in movement, which is what Aristotle yearned for. (Prosaically, I receive ample evidence for the synergy of the mind and the body every time I do intense physical exercise, as a result of which energy is gained the more it is spent and exhaustion gives place to mental lucidity as well as corporeal well-being.) Avowing that the human is a crossroads of spiritual and physical energies; or that spirituality germinates in human interactions; or that it thrives in our non-instrumental treatment of the natural world—avowing all or just one of these assertions, we may refashion the idea of *sunergeia* in the absence of God. How would our politics and economics, natural sciences and cultures be transfigured if they were to rely on synergetic sharing? If extraction and sacrifice were no longer the appropriate course of action? If *enworkment* did not lead to exhaustion and burnout at the macro-planetary and microindividual levels, but to a joyful covibration of differences in sharing?

The Gospel of John holds the potential for a secular resurrection of synergy insofar as it fantasizes about the day when humans would συνεργοὶ γινώμενα τῇ ἀληθείᾳ "become the co-workers of truth" (3 John 1:8). In effect, John's message does not swerve far afield from Paul's: for a Christian, truth (*alētheia*) is so embroiled with Christ that he is *it*, and *the way*, and *the life*. Yet, outside this context, becoming the sunergoi of truth may be understood in terms of a collaboration in the project of un-forgetting—the exact translation of *alētheia*—by sharing energy among finite beings. Or, in a word, living. What un-forgetting saves from obliv-ion is that other energy, which grows more plentiful the more it is set to work with others, and not the least with the other that I am to myself: in the synergy of my mind and body, or of my body and itself, or of my body and various microorganisms that inhabit it. And what synergetic activity doles out to truth is a nonmathematical, nonmetaphysical, counterintui-tive, protocommunist notion of increase (growth) through sharing.

John leads us to the conclusion that truth does not docilely lie in wait for those who would chance upon and uncover it. Alētheia works, sets itself to work, yields works, and actively reposes in itself. His truth is animated by an energy, wherein we are invited to participate, and the name of this energy is *incarnation*. It is up to us to work and rest with and in it. Nothing sways the scales in favor of body or spirit here, since incarnation is the setting-to-work of both, synergetically. Although Christ remains a model for John, *our* good news is that every incarnate being is the enworkment of the living, not amounting to the totality of Life but existing in synergy with others. Biospheres are the works—in the verbal and substantive senses—of countless synergetic ties. To col-laborate with truth is not to have a penchant for spiritual over physical works, for those of "mercy" over those of the flesh. The first precondition for becoming its *sunergoi* is forgetting (more than that: consciously dis-mantling, unlearning) the partitions between body and spirit by way of embracing their elemental synergy.

In order to get to work with others, let alone with the divine Other, one must first find out how to work and to live with oneself. What are the possibilities of cultivating a thinking care for the body and an embod-ied cultivation of thought? How to cope with corporeal, psychic, and spiritual energies concurrently, if some of them block the enworkment of others by dint of their very activation? The concept of entelechy—the soul enlivening a body by assigning it to its appropriate end, *enending*

it—fails to do justice to the image of the human as an intersection of partially overlapping energy fields. More than one entelechy and energy claim us at any given moment, making it necessary to negotiate among them and to let them work in concert, though not as one. All energy is a synergy, the act and actuality of this negotiation; melding Kant and quantum physics, we might say that energy is a synthetic a priori, without which there would be no space-time continuum.

In each of us various settings-to-work are in conflict, giving rise to counter-enworkments (known in psychoanalysis as *resistances*), and so our transactions with others exponentially multiply the opportunities for skirmishes among energies. David Thoreau addresses the quandaries of living with others in the course of explaining his predilection for "solitary dwelling" with regard to the difficulties of establishing interpersonal cooperation: "The only cooperation which is commonly possible is exceedingly partial and superficial. . . . To cooperate, in the highest as well as the lowest sense, means *to get our living together.*"[12] *Cooperation* is the Latin translation of the Greek *synergy*, for which Thoreau reserves a social signification. But "solitary dwelling" is not a panacea from (inner) division, because, even and especially in solitude, one must practice the habit of living together with oneself. It would be wrong to think that operativity and energy are the unitary and coherent processes or products of "individual" existence. They are, rather, internally striated, multiple endeavors forced continually to repel their counterworkings.

Concerning the Aristotelian energeia, we might say that it is a rare quiescence, a cessation of dissension among energies and (a little counterintuitively) an intensification of activity. Its energy of rest—*energiya pokoya*, to use Bibikhin's phrase—is anything but relaxation. The experience of a temporary metaenergetic or synergetic balance that seems to transcend time, it defers the relapse of frictions between enworkments and counter-enworkments, surface and depth, "body" and "soul." In the meanwhile, being unmoved in the plenitude of energeia beckons with the most active of comportments, to the extent that it clears away divisions, resistances, and blockages, thus sanctioning various energies to work together, to put themselves to work together, to be shared, to coexist in synergy, and in this co-operation to become fully themselves.

Spiritual-bodily synergy is an infinite task, with which each can grapple only by her- or himself. Monotheistic religions tend to favor one version or another of Augustine's sacrificial recipe: the works of the flesh are handed over to those of the soul, and both are delivered to God. We have noted, on the contrary, that Hesychasm was committed to embodied spiritual practices with the overarching aim of stilling the breath and the heart so that divine energies would permeate the hesychast. In a programmatic text of synergetic anthropology, Sergey Horujy focuses on the affinities of Eastern Orthodox "synergetics" and Zen Buddhism's "culminating event" of *satori*, a sudden breakthrough or illumination.[13] I would suggest, for my part, testing the synergetic hypothesis of energy's quantitative and qualitative increase in sharing with reference to Ashtanga yoga, for which the *sūtras* of Patañjali, compiled around 400 CE, are a foundational text.[14]

The word *yoga* itself points in the direction of synergy, of putting back together the energies of the body and spirit. Derived from the Sanskrit root *yuj* (*to join, to integrate*, on the basis of which the English *to yoke* is also formed), *yoga*, like re-ligion, means *a binding-together*. The "yoking" that takes place in yoga permits practitioners to achieve embodiment with awareness. *Samadhi*, the deep state of absorbed meditation and concentration attained on the eighth and final rung of yoga leading toward illumination (*Pātañjalayogaśāstra* 3.3), is not a purely intellectual contemplation; it is experienced at the energetic confluence of mindfulness and corporeality, of attention directed to the body, its positions and modulations. The third rung of the ladder (2.46) centers around *asana*, the meditative posture that is to be "steady or stable" (*sthira*) and filled with ease (*sukham*). That is the energy of rest at its most embodied: the body, relaxed and motionless, while working together with the stabilized and attentive mind. The effortlessness of the posture emphasized in the following sūtra (2.47) upholds the quelling of bodily dynamics and dynamism (shifts of position, dislocation, physical agitation, etc.) and shores up a joint actualization of the body and of stabilized mindfulness, tending to the infinite (*ananta, anantyat*). To the uninitiated, it may even sound like the jointure of yoga erases its seams, its "yoke" culminating in the infinite interpenetration of parts that have been thus far divided between and within themselves.

Until almost the very end of the path toward samadhi, the erasure of boundaries between the energies of the body and of the mind remains, nonetheless, incomplete, as breaches (*tachchhidresu*) and possibilities of a sudden turnabout persist in illumination (4.27). The yoke of yoga by no means bolsters undifferentiation and indifference between the elements it interrelates; it only encourages their synergy in working toward enlightenment. Although becoming embodied with awareness alters bodily and mental energies, the resulting energetic whole is, far from a "higher" synthesis of the two, the subtle commerce of differences that, in the final stages of samadhi, "resolve themselves into that out of which they emerged" (4.34). Yogis *descend* to the common source of energies and there savor the liberation (*kaivalyam*) from power and from the power to wield power. Striving to infinity, they aim at energetic concentration in a single point—in space and time, body and mind, perceived as one—that would act as a gateway to infinity. The twist of this aspiration is that theirs is a desireless desire, unfastened from lack, from a craving for something missing. A dreamless energy dream, it must set itself to work *when it is already fully and synergetically at work*—a prerequisite that, apparently impossible to satisfy, is crucial to the practice of yoga. It is in this key that I interpret Patañjali's sūtra 1.1: *atha yoga anushasanam*, "now yoga begins within the (prior) discipline and teaching (of yoga)." In this, too, the energy of yoga accords with Aristotle's energeia. There is no progress per se along its path that entails reactivating the *now* (*atha*), which, like energy itself, commences in the middle of what assumes the shape of a circle. Having delivered themselves to its rotations, yogis are emboldened to remove the obstacles separating them from atha, wherein they intuit the infinite, anantyat.

To overcome the hurdles on the route to liberation, one must unblock the energies that congeal into divisive preferences and attachments, in the first place by making one's thoughts uncolored, *aklishta* (1.5). As a rule, fewer divisions facilitate a greater sharing in synergy, the energies working with, not against, one another. Breath, the synecdoche of energetic sharing between the inside and the outside, is an anchor of my own yoga practice. It is also a bridge between this practice and Hesychasm. The fourth rung of yoga is *pranayama*, the regulation of breath, which grows *sukshmah*: slow, subtle, unmanifest, dormant (2.50). It achieves what Eastern Orthodox spirituality calls *hesychia*, the quietude of the heart and of respiration.[15] The energy of pranayama exceeds that of the circulation of air between the inside and the outside. Nor is it exclusively concentrated in the inner realm, *abhyantara vishaya* (2.51), which more or less overlaps with the heart. Through mindful practice, pranayama disentangles itself from the energy of the depth and of the surface, to be extracted or expanded. In so doing, it flags a transition to the sixth rung of *dharana*, or focus beyond the inner/outer distinction (2.52), itself a precondition for samadhi on the eighth and final rung (3.3) of the circular ladder.

If, in the experience of embodiment, depth belongs to the phase of inhalation and surface to that of exhalation, and if, moreover, the energy of pranayama eclipses both, then this energy's activation depends on the interval between our breathing in and out, the *stambha* where breath is suspended, restrained, and stationary (2.50). Pranayama activates the *energy of the in-between*, of difference imperceptible in everyday life. Mindful embodiment illuminates such overlooked regions of vitality as the disjunction between inhalations and exhalations that border on death, where the pneumatic interval is prolonged indefinitely, or that augur a more intense life. Focusing on actuality at the edge of the possible, it works with ignored spaces and times in the hope of welcoming the enlightenment in them. Yoga discloses the nothingness within and between us: a vital, vibrant, corporeal-spiritual nothingness. As the practice proceeds toward samadhi, it enables us to regain the positivity, the plenitude, of this nothingness. Aristotle's energeia draws near to but ultimately cannot think (or dream) this energy.

3

ECONOMIC CHIMERAS

Energy is at home in economics, fixated on works, and on the workings, of people and machines, or, more subtly, of individual exchanges and commodities, or, again, of financial institutions and the global markets. Economics is an ergonomics—an arrangement, distribution, division, or ordering of ergon. Truth be told, energy is *too* interplaited with an economic sense of work that elbows competing significations out of its semantic range. As I have pointed out, we treat energy as a reserve that could be appropriated and activated as needed, accumulated or wasted, congealed into products or accrued in and of itself as a symbol of power. Within a capitalist framework it exhibits the traits of money, the means for the end of buying "goods and services" that mutates into an open-ended end-in-itself. This analogy is valid and this mutation takes place because the energy of capital finds no actualization in the world it shapes. Even the finished "works" in their capacity as commodities disband into further means of boosting quantitatively measured value. The ensuing conception of energy takes into its ambit nothing but dunamis, translatable into money and power (labor, purchasing, or political) that must keep growing just to secure the positions of those who command them. Human and nonhuman work is denied rest, the finality of achievement, and the actuality of accomplishment. More than that, rest itself is put to work, for instance to boost the entertainment industry, to sustain a broader culture of consumerism, or to prolong "flexible" labor time well into the nonremunerated part of the workday.

In violation of the laws of physics, economic thermodynamics dictates the rules of the game when capitalist energy puts itself to work to the detriment of everything actual. The total value of functioning capital is not conserved but augments, as though by miracle, which Marx links to the exploitation of labor. Its energy field expands, sometimes by virtue of ongoing exchanges, of money changing hands, which is how John Maynard Keynes explains economic growth. As a result of the incessant back and forth between money and commodities, markets can overheat, speeding toward the system's entropy. Then extra-economic interventions, such as increasing interest rates, can help cool them down. Strange forms of life crop up: the dynamic energy of capital dreams of a "city that never sleeps" in contravention of circadian cycles fundamental to all living beings. A city is, to be sure, not a biological organism, but its inhabitants *are*, and they (we) sleep less and less in order to conform to the ideal of beaming energy, ever productive and deprived of rest. Nihilism, boredom, and indifference are the psychological corollaries of this insomniac state steeped in the fear of whatever would impede exchanges, that is, the fear of attachment to anything solid that has not yet melted into air. (As I write these lines, I am not fantasizing about an economic or energetic equilibrium—only of a rhythm that would be more in tune with existence, with life, with difference. At the same time, today's fashionable affirmations of disequilibrium both in the natural world and in human systems might be a thinly veiled ideological device for justifying the shocking and steeply rising inequalities between the richest minority and the poor majority of the global population.)

Capitalist work is as lopsided as the economic energy propagated by this mode of production. Its process and the product, the task and the function, invariably send us back to the "deeper" goal of economic activity, the widening profit margins, or, stated more abstractly using Marx's terminology, the self-augmentation of value. The energy of labor, the material work it animates, and the needs it serves are allotted a secondary role compared to the operations of capital. They are akin to the Augustinian works of the flesh, sacrificed to haughty spiritual works and represented here in a sort of caricature by the ideality of value. Ideologically, the overriding goal of economic activity (namely the production of profits) accords with its source in entrepreneurship, thought to be the engine of production, while labor is demoted to the superficial means for bringing investment strategies to fruition. No wonder that Marx's critique of political economy and its ideological mystifications had to invert the topology of capitalist depth and surface, albeit without foregoing this spatial scheme itself, and to procure in the energy of labor the hidden source of capital.

If we are convinced that capitalist work is distorted and energy misplaced, then how much more so is the idea of economy! Yet another word of Greek pedigree, *oikonomia* is a combination of dwelling (*oikos*) and law, division, distribution, arrangement, or order (*nomos*). Capitalist economy, however, is the law (of value, of supply and demand, or of the market) minus the dwelling, decontextualized and uprooted from local belonging. Since its dawn on the ruins of European feudalism, it has been spreading global homelessness, dislodging entire populations and social subgroups. At the same time, "homeworking" has persisted in the guise of unpaid, mostly female, labor performed in the house and, recently, in the trend of working from home made possible by the new means of communication. We can study these and other empirical economic realities sociologically or anthropologically, but, to address them philosophically, it behooves us to examine the tensions within capitalist eco-nomy, where the energy of displacement propels the circuits of exchange as well as the early industrial relocation and postindustrial "flexibilization" of the labor force.

———— ∞∞∞ ————

Capitalist economies undermine themselves by implacably executing their law, depriving us of the dwelling at every level, from individual household to planetary abode. The self-undermining I am getting at is not the internal contradiction of *capitalist* economy enunciated by Marx, i.e., the falling rates of profit due to the gradual replacement of human labor by machinery. I am referring, instead, to the contradictions plaguing capitalist *economy*, in which the law impinges on the dwelling and transforms it into a space inhospitable to life. The organizing nomos wipes out the very oikos it proposes to organize and thereby descends into anomie. The environmental crisis is one facet of eco-nomy's inner fissuring that prioritizes energy extraction and accumulation in the form of "natural resources," uncompensated products of human labor, or plant and animal growth.

Actually existing socialisms displayed a comparable disregard and damaging attitude toward the environment. Unrelated to deficiencies in the know-how or inadequacies in state-of-the-art "clean technologies," their predominant attitude failed to reconcile the two constituents of eco-nomy, to work at and on work, to question the tyranny of productivism. The economic energy of actually (the word 'actually' in this expression is highly misleading: first, because no socialisms exist in our actuality and, second, because in this context the word both names and occludes the energy of these economic regimes) existing socialisms retained most capitalist markers, except that the centralized state apparatus accumulated it in place of privately owned corporations. Under Stalinism the energy of displacement was booming also in the Soviet Union, at the cost of millions of lives. And the housing crisis, particularly in big cities like Moscow, made dwellings scarce and dwelling precarious, with neighbors spying on neighbors in the infamous communal apartments, one of which I inhabited as a child.

To dream of another energy of dwelling and working is not, by default, to bemoan the forfeiture of a parochial and "organic" relation to one's surroundings. Capitalism is indeed the economic force of liberating negativity that dissolves the suffocating ties of feudalism, oppressive communities, antiquated gender relations, and so forth. But its circuits are untenable in the long term, dependent as they are on the dunamis of quantitative value that undercuts actuality, negating the materiality of the producers and that of the "externalities," to which it relegates the environment as a whole. Its environmental unsustainability is an upshot of the unsustainable potentiality that melts, vaporizes, or burns through the actual. Never finished, the accumulation of capital (value's self-valorization) is the inverse image of Aristotle's energeia, of thought thinking itself.

On the face of it, ancient Greeks denounce the abyss between the noble activity of *theoria* and the plenitude of thinking energy, on the one hand,

and base economic pursuits, on the other. The realm of freedom from material concerns and from external determinations in the former is an antipode to the realm of necessity, allowing us to secure the minimal conditions of bodily existence in the latter. More shamefully still, the Greek oikos, in the context of which oikonomia first took shape, was a household of slave owners, assigned rigidly divided gender roles and functions. But this should not detract from texts such as Xenophon's *Oeconomicus* or Aristotle's *Oeconomica* that reveal to us the forgotten energy of dwelling.

Economic law, in Aristotle's treatise, is grounded on sexual difference between the husband and the wife, the relationship it is commissioned to regulate and upon which it simultaneously depends. The heterosexual couple is the nucleus of the household (*oikia*), made of the human component (*anthropos*) and of chattels or goods (*ktesis*) (*Oecon.* 1343a, 15–20). The kernel of economics, then, is not a single individual. The nucleus parts into two, who "form a community out of necessity [ἐξ ἀνάγκης αὐτῶν ἡ κοινωνία συνέστηκεν]" (1343b, 10–15). Crucially, the energy of the couple, however tightly regulated its gendering and composition, is a synergy that puts the house to work in a specifically human vein, with the view to the common good promoted by "mutual help and good will." Although it arises out of necessity, the fundamental economic community is the "synergy between woman and man aiming not merely at existence but at a happy existence [ὅτι οὐ μόνον τοῦ εἶναι ἀλλὰ καὶ τοῦ εὖ εἶναι συνεργὰ]" (1343b, 15–20). Its synergy is the enworkment of well-being (or "being well," εὖ εἶναι).

Oriented toward the substantive end of the good—the aim of all human activities—economic life according to Aristotle is unconcerned with the accumulation of profits or wealth for its own sake, with the expansion of the household and its possessions, or with the efficiency with which the oikos functions. The energy that puts the common dwelling to work is attuned to an order that would be just, which is to say, fine-tuned and respectful of the multiplicity it articulates. Aristotle worries about the justness of this order-nomos, not the least in his defense of agriculture as "the most honest of occupations [ἡ δὲ γεωργικὴ μάλιστα ὅτι δικαία], since the wealth it brings is not derived from other men" (1343a, 25–30). We might criticize his allotment of justice to human appropriators alone, a decision that leaves plants and animals vulnerable to the proprietary drive, just as we might vent our righteous anger at the circumscription of gendered duties to men defending the house and women keeping watch over it (1344a, 1–5). But, regardless of the epochal limitations that hew the texts of every thinker, the pervasive concern for justice and the good we chance upon in Aristotle recasts the energy of economy in a light other than that of the means without end propagated around us. Its beginnings steeped in necessity, the community (*koinonia*) that sets itself to work for the common good, in synergy, will reverberate across the millennia in Marx's communism. The other energy of economy, then, is ethical.

Ethics and the good are allied, in the world of ancient Greece, to being. Beings exist by virtue of their preservation (in being), which saves them from encroaching decay. And that is good. To be is to be kept safe in being; the ethical energy of economy is, ipso facto, ontological. Along these lines, Aristotle enumerates the "four aspects of an economic actor [Εἴδη δὲ τοῦ οἰκονόμου τέτταρα]": coming to possess, preserving, improving that which exists, and making use of it (1344b, 20–30). Each type of economic action is reminiscent of a relation to being, by virtue of putting to work—or actuating—things and assigning them to their appropriate ends. In our economies, in contrast, only possession survives. Preservation and qualitative improvement are out of the question. For all the clamor of pragmatism, "making use," *chrēstikós*, is hopelessly opaque to us. It does not mean grasping something as a mere means and utilizing it as an instrument for an external goal. Quite the contrary, *chrēstikós* calls for the dispensation of a thing to the necessity (*chrē*) proper to it, its essence actualized in existence. The Aristotelian economy is care for beings according to their being, guided from what they are in potentia to their enacted liberation as and for themselves, the enworkment of their ends. Reunited with their essence carried through to actuality, they energetically rest in it and are preserved. With respect to property, as well, good economic management channels necessity into freedom, thanks to the fourfold activation of the things' ontological ends. The same is applicable, mutatis mutandis, to slaves, who, following the demand of justice (*díkaion*), are to be freed after a limited time of service (1344b, 15–20). A good ordering of a dwelling, oikonomia, pivots not on a despotic and arbitrary will of the estate owner but on the energizing of being helped along its way from potentiality to actuality.

The ontological subtext of capital is that, when the abstract freedom of value forges abstract equivalences among heterogeneities, the concrete freedom of energetic self-accomplishment suffers. It is not that, in the absence of value, things and people are enclosed in themselves, monadic, and self-referential. Incommensurable with one another, their actuations and actualizations are intertwined in the common place of dwelling. Once economy is mutilated as the anomic nomos bereft of oikos, the universal relations of value take over this absent spatial, embedded, singular milieu and put everything and everyone—everyone *as* everything—to work for the same extraneous end without end of a self-expanding potentiality. A dull imitation of existence with no existential richness and diversity, an economy without an oikos drains the energy of its human and nonhuman participants into an effective but nonactual totality of capital. Collecting the types of *Entfremdung* (alienation) Marx detailed under another heading, we may thus posit an *ontological-energetic alienation* of being from itself, of its potentiality from actuality, the estrangement that diverts its enworkment to an alien growing power parasitically fed by its frustrated possibilities.

As for Aristotle's treatise, it systematizes the arguments of the Socratic dialogue *Oeconomicus*, recorded by Xenophon. In the text's incipit, Socrates puts the energy of economy in question, inquiring into the work, ergon, it carries out (1.1). By way of responding, his interlocutor, Critobulus, shifts the topic to ethical grounds: the assignment of "a good economic actor is to make a good dwelling out of his own dwelling [οἰκονόμου ἀγαθοῦ εἶναι εὖ οἰκεῖν τὸν ἑαυτοῦ οἶκον]" (1.2). The reiteration of *dwelling*, preceded first by the qualifier "good" and then by "his own," is not fortuitous; Critobulus hints that economic work worthy of the name actualizes the potentiality of a formally owned dwelling to be a household—the potential of possession to serve the good. This inaugural repetition, which touches on the being of the dwelling energized by and for the good, is more suggestive than the philosophical questioning the word will undergo at the prompting of Socrates (in 1.5). It conveys that a good "economist" refrains from dominating the dwelling and develops its energy in the best way appropriate to the oikos. Though unmentioned here, the nomos of economy is a conduit that channels the energy of the dwelling toward its inherent end, the good.

Irreducible to an afterthought or an ornamentally superfluous addition, the ethical spotlight on economic affairs is vital for their functioning. The oikos of oikonomia holds together possessions that uphold the good; it embraces "anything good that he [the owner] possesses [ὅ τι γε τις ἀγαθὸν κέκτηται]; of course, I don't call anything bad that he may possess properly" (1.7). In English we espy the ethical footing of economy when we designate property as *the goods*. We also ask about the end or purpose of something, *What good is it?* In raising this question, we do not assume there is one supreme Good that would accommodate the miscellaneous things or activities we are inquiring about. Each time unique, the good is ingrained in the goods, and the mission of a skillful manager is to find out how to set it to work, releasing its energy, liberating it, making it free to become the good that it is. (Again, in English, the good may amount to actuality or actualization: to fulfill a promise is *to make good on* it.) The anticipatory definition of economy that stems from these reflections is that it is the energy of the good, that is to say, the good—or the goods—put to work for the good. The energy of the dwelling and its law are secondary to that of the good, which goads them on, articulates them, lets them live up to *their* promise, makes good on them.

More concretely, Socrates believes—and in this Aristotle follows him to the letter—the "work of agriculture," *geōrgía ergon*, to be the best type of economic activity, blessed by the gods (5.19) and conducive to a pedagogic synergy (*sumpaideu*) of training in virtue (5.14). Every direct appeal to energeia in the text orbits cultivation, "the earth that is put to work," ἐνεργὸν οὖσαν τὴν γῆν (4.8), ἡ γῆ ἐνεργὸς ἔσται (4.9). It is as though agriculture energizes the earth, enworking it such that it would be helped along to its own innermost good, a fertile actuality. Certainly, the fundamental economic energy Socrates conjures here would not apply to overfarming or to "intensive agriculture," which puts the earth to work while forgetting its good. Overfarming disregards the inner measure of the earth, a geo-dunamis that comes to fruition in energeia and undergirds the nomos of economy.[1]

To gauge the inner measure of things that dictates the law is to return to the things themselves, to learn from them, and to learn, too, from the earth: "because the earth is a goddess she also teaches justice to those who can learn; for the better she is attended to, the greater the good she does in return [ἔτι δὲ ἡ γῆ θεὸς οὖσα τοὺς δυναμένους καταμανθάνειν καὶ δικαιοσύνην διδάσκει: τοὺς γὰρ ἄριστα θεραπεύοντας αὐτὴν πλεῖστα ἀγαθὰ ἀντιποιεῖ]" (5.12). The justice taught by goddess-earth is the very inner measure that is codified in a good law. According to this insignia of finitude, "those who can learn" care (proponents of biodynamic—but why privilege dunamis here?—agriculture will be pleased to know that the Greek verb to which Socrates resorts, *therapeuontas*, calls for a therapy) for the earth's potentialities in the best possible manner, patiently attuned to its capacities and limitations. Agricultural and, more broadly, economic activities succeed by virtue of their attunement to what they work on or with, enhancing the materials' own ontological enworkment, bringing their energy out into the open. The goods that come out of the earth, whose potentialities are respected, are the actual products, the erga, of its good set to work independently of human beings and released into its own, in a send-off prepared by our caring, therapeutic attention.

To most of us, the idea of the inner measure of things sounds impenetrable at worst and conservative at best. We have been raised on the fantasy of limitless achievement, the unconditional merit of overcoming previous constraints putting us on the energy treadmill. Respect for the inherent limits of the human body, of economic production and consumption, or of the earth's capacity to sustain our industries is taken, with utmost suspicion, to be a sign of complacency. Thriving on the production and reproduction of surplus, our economies multiply goods without ends and without end, goods without the good, oozing an unbound derealizing energy that spills over into a culture of obsolescence. Work is dissociated from the care for and attunement to the potentialities of the worked-upon. It is a mechanical expenditure of energy with the activation of a mechanism, indifferent to the materials it processes, for a prototype. The market, where so long as there is demand anything can be bought including human labor and student or credit card debt, is the institutional apex of this indifference.

In truth, market 'neutrality' only exacerbates that of money, which greases the wheels of economic transactions in nonbarter systems. It does not matter in the least what commodities are equalized through the prices expressed in monetary terms. The question is whether money itself has energy and, if it does, then of what kind.

Within the scheme of economic energy, money is pure dunamis, a potentiality neither actualized nor actualizable in any conversion into "goods and services." This trait it has in common with power. The means missing a clear end, it is anathema to Greek thought. Aristotle, for one, lambastes the life of moneymaking (*chrematistikē*) as unlimited (*apeiron*), authorizing no completion, and absorbing "all the energies in the business of getting wealth" (*Pol.* 1258a, 1–5). Money procured for the sake of money is a vortex, replete with a negative "energy" that thwarts, by indefinitely deferring it, the enworkment of the good. It drains activities of their proper ends as soon as it interjects itself into our lives instigating an endless pursuit. Aristotle will go on to criticize usury, or money begetting more money (or "the interest increasing itself," ὁ δὲ τόκος αὐτὸ ποιεῖ πλέον), as "being against the nature of that which money is for" (*Pol.* 1258b, 5–10).

The Greek word for financial interest, *tokos*, is homonymous with "offspring." "As product," Derrida explains, "the *tokos* is the child, the human or animal brood, as well as the fruits of the seed sown in the field, and the interest on a capital investment: it is a *return* or *revenue*."[2] Produced as a living work, it emanates from the generative energy that comes back to itself with more of the same. Aristotle's rejoinder to the economic imitation of growth and reproduction is that, in contradistinction to a human, an animal, or a plant, money is a (technical) means that, *per definitionem*, has no actuality other than to become other to itself in the value of the object purchased with it: "money came into being for the purpose of exchange [μεταβολῆς γὰρ ἐγένετο χάριν]" (*Pol.* 1258b, 1–5). The energy it commands is borrowed from another source, from the external end it must serve. Tokos is a kidnapped, stolen child.

Marx updates the Aristotelian theory, germane to the formula of financial capital, in volume 2 of his *Das Kapital*. Money begetting more money
(M-M') is an abbreviation of the "normal" circuit of capital that swells
in direct proportion to the exploitation of wage labor: "It goes without
saying, therefore, that the formula for the circuit of money capital . . . is
the self-evident form of the circuit of capital only on the basis of already
developed capitalist production [*selbstverständliche Form des Kapitalkrei-
slaufs nur auf Grundlage schon entwickelter kapitalistischer Produktion
ist*], because it presupposed the availability of the class of wage-laborers
in sufficient numbers throughout the society."[3] The abbreviation of financial capital is a second-order obfuscation that further veils the mystified
exploitative relations of production founded upon the extraction of surplus value from labor. Hence the energy behind M-M' emanates not from
money itself but from the wage laborers forced to give up a fraction of
their labor power, uncompensated. Put to work in a bank or in a hedge
fund, money works through others and by other means, powered by an
energy supply that lies buried in the "deep abode" of capitalist production, as Marx is wont to refer to it. Financial capital is a synecdoche of
capital in general that also occludes, if to a lesser extent, its provenance
from the nonrestituted share of the workers' energy.

Evidently, the valuation of the inner over the outer survives nearly unal-
tered in Marxist political economy. Kept in reserve in labor power (itself
only a capacity to work) as the workers' dunamis and their sole "possession," energy is extracted from them with the unstated threat of violence,
the silent menace of starvation and death that await those without a paid
job. To add insult to injury, capitalist blackmail is then wrapped in the
ideology of freedom, according to which one is free to sell one's labor
to any interested employer or to die without any means of subsistence
whatsoever. The sphere of economic circulation, be it of commodities or
of money, presents the facade of an independent energy of the surface,
the dream of the market's self-activation, movement, or march propelled
by its autonomous laws of supply and demand. Marx's *critique* of political
economy is unequivocal in this respect: *free market* is an ideological fiction that beclouds its origins from the energy of depth, extracted by the
mechanisms of class exploitation. Economic circulation coordinates the

flows of energy both in the subjective-abstract form of money and in the objective-concrete form of products. It overlays the subjective-concrete energy of the workers activating capital, unconsciously giving it actual existence, yet without having the right to decide on how to put it to work. The further away one is situated from this furnace of economic energy, the less discernible its workings.

It was not until 1935, the year Keynes published *The General Theory of Employment, Interest, and Money*, that an independent energy of economic surface became plausible. Even though analogies between phenomena in social and physical sciences may be somewhat jarring and problematic, I submit to readers the following comparison by way of a heuristic device: Marx's political economy is to Keynes's "general theory" what Newtonian physics is to Einstein's theory of relativity. Newtonian mechanics separates an object's energy from its mass. An object at rest has no kinetic energy, and any other energies it contains (chemical, thermal, etc.) make no proportional contribution to its mass. Einstein's equation $E = mc^2$ conversely posits a mass-energy equivalence in relation to the velocity of light, squared. Presumably resting, particles have a rest-mass energy, not a posteriori added to the object but of a piece with it. And when their movement approaches that of the speed of light, their mass grows exponentially, no longer equivalent to molecular weight at rest.

The velocities of circulation are, analogously, in Keynesian economics, imbricated with the production and reproduction of value. The levels of financial liquidity vary depending on the willingness or "unwillingness of those who possess money to part with their liquid control over it" and to be rewarded according to the interest rates at the time.[4] If interest rates are high, there will be few incentives to 1. maintain high liquidity or 2. invest in anything other than financial capital. "It seems," Keynes concludes, "that the *rate of interest on money* plays a peculiar part in setting a limit to the level of employment, since it sets a standard to which the marginal efficiency of a capital-asset must attain if it is to be newly produced."[5] The brakes on and accelerations of money circulation and of financial markets (in a word: the energy of the surface with all its unevenness) influence the levels of production and employment where the classical energy of depth has originated.

Economic energy and mass are interrelated through the velocity of monetary circulation, which supplants the speed of light in Einstein's equation. In *The Treatise on Money* Keynes qualifies the imparity between income velocity and transaction velocity,[6] corresponding to the speeds of movement at the core and on the superficies of the economic sphere. The surface of transactions needs to be factored into the calculation of the "true velocities" of cash flows that account for the differences between the "income-velocity of money" and "accumulated savings,"[7] that is, in the language of Einsteinian physics, between relativistic mass and rest mass.

Once the energy of circulation comes to the forefront of the economy, the latter is cut loose from the enworkment of and for the good. The bedrock of production was once the raison d'être of economic activity, busy with the fabrication of useful works, or, if not altogether useful, then those for which there was effective demand. That old energetic scheme was relatively uncomplicated. The laborer's work (menial or intellectual, the exertion of muscles or of nerves, as John Stuart Mill put it) was the subjective side of energy,[8] which, at the end of the production process, was objectified in the works, the finished products. Further along, we will reflect on how the workings of capital have upset that economic energy correlation. What is of interest here is that, when the fickle velocities of the monetary mass come to regulate employment levels, the "borrowed" energy of money becomes primary, the supplement preceding that which it supplements. The significance of production and its goals fades; low interest rates animate investments outside the financial sector regardless of the substantive ends of the economy, i.e., the goods to be produced and the good to be drawn from them. Conceivably, intense economic activity would continue undiminished in a climate where interest rates are kept low, including in a hypothetical situation of abundance where effective demand is saturated. Upon isolating income velocities from transaction velocities, the gross domestic product (GDP) will grow in equal measure when the same money changes hands as when more money is spent by workers with rising incomes. And therefore productive and consumptive energies alike will be, at least theoretically, utterly expendable.

Economic policies gathered under the umbrella of Keynesianism have marshaled the energy of the surface, divorced from the question of the substantive good and freed for an ever accelerating circulation. They have followed to the letter the advice for government investment that Keynes himself doused with a great deal of irony:

> If the Treasury were to fill old bottles with banknotes, bury them at suitable depths in disused coalmines which are then filled up to the surface with town rubbish, and leave it to private enterprise on well-tried principle of *laissez-faire* to dig the notes up again . . . the real income of the community, and its capital wealth also, would probably become a good deal greater than it actually is. It would, indeed, be more sensible to build houses and the like; but if there are political and practical difficulties in the way of this, the above would be better than nothing.[9]

Seemingly inane government investment in airports servicing a village or two, bridges to nowhere, or several parallel highways single-mindedly insists on what is good for the economic system itself: quantitative growth. Quite beside the point are the questions regarding what grows, how, with what impact (environmental or otherwise), and for whom. At the surface, energy increases for energy's sake, or, more precisely, only for the sake of increasing. We ought to stop ignoring this expanding energetic superficiality as though it were an unfortunate by-product of capitalism and insist on reconnecting it to the energy of depth until the rigid barriers between the two are undone, in economic and other branches of human activity.

⎯⎯ ⊗⊗⊗ ⎯⎯

For a long time classical political economy was readying the theoretical infrastructure for the Keynesian energy revolution. Adam Smith made "the wealth of nations" the centerpiece of his inquiry, as did John Stuart Mill, who proclaimed that the proper subject of political economy was wealth, comprehended not just as money but as anything "which has a *power* of purchasing; for which anything useful or agreeable would be given in exchange."[10] I italicize the word *power* to show how, with the monetary or monetarist overtones inflecting both consumption and production, it transports the energy of the economic realm to the territory of potentiality—and an empty potentiality at that—without the actual in sight. The trasport in question is no small feat: it is an egression from the ethical underpinnings of economy toward its modern political bases. Now, if Marx's mature project is framed in terms of a "critique of *political* economy," as the subtitle of *Das Kapital* indicates, then what it really criticizes, besides the unequal distribution of power (read: dunamis) between the increasingly polarized classes, is the non-fulfillment, the betrayal of human self-actualization through labor, and, at bottom, the exclusion of energeia from the field of capitalist economic activity.

Marx's assessment of the wealth generated under capitalism in the "Economic and Philosophic Manuscripts of 1844" is a damning one: "The worker becomes all the poorer the more wealth he produces, the more his production increases in power and range."[11] The realized potentiality of labor—the power and range of production—does not attain to actuality due to the interference of the capitalist relations of production with the workers' self-actualization. The energy of labor is expended so that its putting-to-work works against the workers and, in the long run, puts more and more of them out of work, according to the falling rate of return and the replacement of the human workforce with machinery.

When we study Marx through the prism of energy, we perceive no epistemic break, which Louis Althusser emphatically affirmed, between his "early" humanist, and "late" antihumanist texts. An unremitting worry about alienation that suffuses Marx's corpus is given numerous outlets into concerns with the labor process, with the product of labor, with human self-alienation . . . But the common denominator of these apprehensions is the slipping away of energy, now in the subjective form of productive activity, now in the objective form of the finished product, and now in the form of self-actualizing labor, which combines the two. Marx dreams of maintaining all attributes of energy together, of preserving its wholeness, of not letting it drift away, depart too far from us and, especially, from itself, of preventing it from parting with and against itself. To claim that estranged labor, *die entfremdete Arbeit, is* capital is to say that it is energy siphoned away and deployed in an abstract ruse against those who gave it actual existence.

It is high time we read Marx's texts energetically, with renewed energy and placing the stress on a theory of economic energy simmering between their lines. For instance:

> Till now we have been considering the estrangement, the alienation of the worker only in one of its aspects, i.e., the worker's *relationship to the products of his labor.* But the estrangement is manifested not only in the result but in the *act of production—*within the *producing activity* itself. . . . The product is after all but the summary of the activity of production. If then the product of labor is alienation, production itself must be active alienation, the alienation of activity, the activity of alienation [*so muß die Produktion selbst die tätige Entäußerung, die Entäußerung der Tätigkeit, die Tätigkeit der Entäußerung sein*].[12]

The dual sense of *ergon* (working and a work) migrates from Aristotle's thought, via Hegel, to Marx's writings. What has been put to work and yielded results is taken away from the workers, in the degree that their putting-to-work itself works against them, deprives them of "wealth," of the process as much as the product, of energy. "Active alienation, the alienation of activity, the activity of alienation" is energy counteracting itself in the course of its enworkment, undermining and annihilating itself in more manners than we can imagine. Under these conditions, exacerbated by a minute division of labor and wild competition, there is no synergy, no working-with or with-work, which means that there is no viable energy, either. The crises of capital are acute manifestations of this nonexistent synergy, of the energy imbalances upon which capital is built.

Capital interjected between the worker and the work, between workers and the means of production, and within the labor process splinters ergon, forbidding the gathering of energy into an actuality. To the question "What then constitutes the alienation of labor?" Marx answers: "First, the fact that labor is *external* [*äußerlich*] to the worker . . . ; that in his work, therefore, he does not affirm himself but denies himself, does not feel content but unhappy, does not freely develop his physical and mental energy but mortifies his body and ruins his mind [*nicht bejaht, sondern verneint, nicht wohl, sondern unglücklich fühlt, keine freie physische und geistige Energie entwickelt, sondern seine Physis abkasteit und seinen Geist ruiniert*].[13] Activated (enworked) from the exteriority of capital investment, workers continue to work outside themselves, their energy beside itself, deranged. Their self-negation and "unhappiness" are the traces of energeia's retrocession. They are dispossessed of the energy of rest, which is emphatically not a sign of indolence but is concomitant with the free development, inner enworkment, and outward expression of finite being.

The bid, launched by Marx, to right these wrongs is ambitious in scope. He wishes nothing more and nothing less than to resolve the crisis of energy by reconciling the verbal and substantive aspects of its work, or enworkment, through the workers' self-appropriation with respect to the process and product of their labor. Communism names the synergy of this self-appropriation in common, which, for the first time, will have given an outlet to human energies and inaugurated human history after a long phase of prehistory. Hegelian actuality (Wirklichkeit, being the retrieval of Aristotle's energeia) is the label Marx affixes to communist action, the workers putting themselves to work, activating themselves, politically and economically, politically-economically, in the plenitude of the energy proper to them, which is already a synergy and therefore im-proper, not appropriable by any single individual: "In order to abolish the *idea* of private property, the *idea* of communism is completely sufficient. It takes *actual* communist action to abolish actual private property [*Um das* wirkliche *Privateigentum aufzuheben, dazu gehört eine wirkliche kommunistische Aktion*]. History will come to it; and this movement, which *in theory* we already know to be a self-transcending movement, will constitute *in actuality* [in der Wirklichkeit] a very severe and protracted process."[14] Ordinarily read as evidence of Marx's historical determinism, this proclamation is a manifesto, a "performative" act, a dream of more-than-an-idea (which Kant had earlier entertained, as we shall discover, with reference to perpetual peace) that would open the sole avenue for the flourishing of energy through its self-determination, self-actualization, or self-activation carried out in common, synergetically. The caveat is that communist energy must be actualized before it is actual, prior to its coming to fruition in collective action. *Eine wirkliche kommunistische Aktion*, an actual communist action is pure, circular energeia—a self-actualizing actuality. Political-economic work has to be carried out before it begins. To be won, class struggle must be indexed to energetic rest.

We have touched upon the twinned political and economic limits of political economy at a point where the actuality of action escapes the system of productivism. It will be of vital import to forge an association, a synergetic bond, between these limits and a community that is, as Jean-Luc Nancy put it, *inoperative,* and that is sympathetic to Georges Sorel's proletarian general strike with its suspension of "productive" work, which reproduces the worker's existence strictly as an appendage to the reproduction of capital. I will maintain, in chapter 5, that inoperative communities and general strikes are *more* energetic than the productive business as usual, not less. They instantiate the positive powerlessness of accomplishment, putting an end to work that operates on behalf of capital and that follows the predictable pattern of means and ends. Does this powerlessness mutate, in tandem with Nietzsche's critique of Christianity, into a new power, for instance when carried through to the workers' physical death, as in Sorel's general strike? What about the fancies of self-actualization the Italian *Autonomia* movement for labor emancipation sustains? These are political questions, and we will have to wrestle with them on a later occasion. Let us turn our attention, in the meantime, to inoperativity and nonproductivity that, from within the economy *proper*, challenge the economic harnessing of energy.

The bifurcation of capitalist topography into labor and capital, itself a potentiality of the workers diverted from them, renders various meanings of production incommensurable with one another. That which is productive for labor may be unproductive for capital. A productive worker is someone whose "production is taken over by capital and only

occurs in order to increase it. A singer who sings like a bird is an unproductive worker. If she sells her song for money, she is to that extent a
wage-laborer or merchant. But if the same singer is engaged by an entre-
preneur [*von einem Unternehmer engagiert*] who makes her sing to make
money, then she becomes a productive worker, since she *produces* capital
directly."[15] Unproductive labor relishes its energy, its enworkment, for an
end tied to the materiality of the labor process or, perhaps, for no end at
all. It fuses with play. Productive labor is activated, "engaged," enworked
for the sake of capital accumulation. Regardless of the volume of its
material or immaterial output, unproductive labor will drop out of capitalist accounting and will not count for anything so long as it does not
contribute, more or less directly, to the self-valorization of value. It will
be tabulated together with economic energy and opportunity lost while
being regained by the workers themselves. To ensure itself against this
possibility, to get a hold of and channel energy into productive outlets,
capital interposes "entrepreneurship" between the worker and the work,
seizing this "between" (*entre-prendre*), under-taking (*unter-nehmen*) its
activation, the engagement of its energy for the sake of someone or something other than the worker. The spiral of capital's insatiable dunamis
shatters the circle of proletarian energeia. Capitalist production disrupts
the work of labor, which it deems unproductive on its own terms.

    What is produced in this spiral is, consequently, quite unimportant;
"labor with *the same content* can be either productive or unproductive."[16]
Rather than content, the formal production of the producers themselves
makes all the difference in a starkly divided world where the energy
dreams of capital are the nightmares of labor and of the earth. Once
Marx reconnects the economy to the good in the simplest sense by ask-
ing "*Who* is this good for?", he is able to demonstrate that the good is not
one and that it is separated from the actual goods, the use-values, that
result from economic activity. But, unfortunately, his black-and-white
portrait of economics excludes the gray area of self-employment ("she
sells her song for money")—a self-engagement or self-activation that
disrespects the lines of division, drawn by capital itself, into the seg-
ments productive and unproductive of capital. Is a self-employed singer
or worker exploiting herself, standing in relation to herself in the shoes of

a capitalist *and* a laborer? Is her enworkment external and alienated, in that its main purpose is to make money? Or does she recoup something of the Aristotelian energeia and happiness? Doesn't the market, in any event, erase the productive origin of the commodities bought and sold there? In this hesitation between the inside and the outside, between self-activation and an alien motivation, another conception of energy announces itself outside the closed system of production and its shadow, unproductive waste.

Just as Marx denies a constitutive role to the energy of the surface generated by the circulation of money and commodities, so he turns his back on an economy that is creative beyond the productivist imperative, which includes, ideally, the workers' deliberate self-production. In capital's semantic regime, extra-economic realities are the not-yet-commodified territories to be conquered by the potentially total expansion of self-valorizing value. Reducing energy to production and antiproduction, Marx reifies economic activity otherwise, by leaving out those of its effects that do not yield a product—including, above all, oneself, one's actualized "human essence"—but inhere within the open-ended process of work. A work that produces nothing, or no thing, be it as immaterial as a service, is no longer recognizable qua work. Its energy delights in itself alone, is gratuitous, expended for nothing, to put it in the terms I am adopting from Georges Bataille's *general economy*. It is in the unproductive sectors of the economy that, according to Bataille, we can see the economy's dependence "on the circulation of energy on the earth [*parcours de l'énergie sur le globe terrestre*]" and so pose economic problems as "*general* problems that are linked to the movement of [this] energy."[17] Besides the sort of work that is useless for the growth of capital, we might isolate a *generally unproductive* activity (Marx's singer who freely "sings like a bird")—the motor of any economy and of energy itself. In this case, enworkment and employment will have the overtones of "emplayment": putting into play what, in some instances, will turn out to be work. Employment and emplayment converge on the energetic horizon of nonalienated labor.

Bataille tracks economic energy unburdened of productivism back to the excess of life, uncontainable in a living being. "The living organism," he writes, "in a situation determined by the plays of energy on the surface of the globe [*les jeux de l'énergie à la surface du globe*] ordinarily receives more energy than is necessary for maintaining life; the excess of energy (wealth) can be used for the growth of a system (e.g., an organism); if the system can no longer grow or if the excess cannot be completely absorbed into its growth, it must necessarily be lost without profit; it must be spent, willingly or not, gloriously or catastrophically" (21). The play of energy Bataille ferrets out is markedly superficial in its whirling on "the surface of the globe." It belongs to the exteriority of an organism or a system unable to absorb everything it receives from the world and dispensing the excess back to its other, "gloriously or catastrophically." The best and the worst overlap in the gift excised from the networks of exchange. To tap into this energy source, emplayed-enworked in life itself, one need not extract it from the atom, from the potentialities of "raw materials," or from labor power-dunamis. One only ought to engage with the excess of that which cannot be contained within the system, with the place, utopian on the system's terms, where work and play mingle in the energy of rest. Such would be, also, a more charitable reading of Marx's "nonestranged labor"—the intersection of enwork-ment and employment in a plenitude uncontainable within itself.

Alluring as Bataille's excessive and superficial energy may be, it has the feel of an overreaction to the classical fixation on production and depth. The expenditure of energy without return is a work-play with-out works, forgetful of the substantive, objective implications of ergon. Totally exogenous, it is a spatial and temporal *ex*ergy, rather than energy, the squandering of mere consumption, an extravagant spending that leaves no room and no time for energetic rest. In reality, aimless expen-diture sidelines production without overcoming its purposiveness. Its dynamics are consistent with the global economic imbalances between the consumption-oriented societies of the North and industrial pro-duction concentrated in the South. Presented with Bataille's argument, I begin to find the crassest version of Marxist materialism attractive: it is not consciousness that determines being, but being that determines consciousness, and so the thinking of energy as superfluous waste could have originated nowhere but in a wasteful and luxurious mode of being characteristic of the postindustrial lifestyle. The sentence "I insist on the fact that there is generally no growth but only a luxurious squandering of energy [*une luxueuse dilapidation d'énergie*] in every form!" (33) makes sense in a milieu of satiety, not of hunger. Bataille's living organisms and economic systems dither between dearth and surplus; howbeit, most frequently, they veer toward the *too much* that overflows the constraints of the organismic and systemic capacities to receive and to accumulate.

Nor does Bataille part ways with capitalist ideology that segregates economic activity from the good. If anything, he exacerbates this tendency. The "overabundance of energy [*la surabondance de l'énergie*] on the surface of the globe" is attributed to solar radiation, mostly refracted back by living matter in a useless expenditure of what remains after some of it has been converted into growth (29). Two issues hang in the balance in this assertion: utility (or lack thereof) and the naturalization of value.

Utility. Whereas the Sun of metaphysics was analogous to the Good, the excess of actual solar light and heat in general economy is beyond good and evil. But what appears to be useless, good for nothing, may be a harbinger of another use, a player in another economy. Spinoza was right: contingency is a side effect of our warped and incomplete knowledge of necessity. Uselessness may be flipped around and become supreme use, just as powerlessness can attain the peaks of the highest power. At the bare minimum, expenditure without reserve is useful in that it cleanses the system or the organism of the excess it cannot handle in any other way, lightening the burden, as it were. On another plane, it obeys the demands of a sacrificial offering (the latest epitome of which is "the society of consumption," Bataille confesses) that plugs human work into the energy of divine economy (45). Energy can be translated into growth directly or obliquely, positively or negatively, through the apparently pointless squandering and waste. We must beware of reducing this energetic mode to pure expenditure, the error that forms the backbone of Bataille's thought.

Naturalization. If the sun is the source of energy and its excess, then the prototype of general economy is agrarian. Preindustrial at the level of production, this economy is postindustrial in the sphere of consumption. Bataille, therefore, risks incurring the same ironic criticisms as those Smith and Marx leveled against the Physiocrats with their conviction that value grew on trees. In particular, he flirts with the naturalization of surplus value—a historically conditioned excess distilled from variegated methods to exploit labor—elevating it to an exemplar of a cosmic, universal phenomenon. Spliced into the exorbitant circulation of biochemical and solar energies, economic structures present themselves in the guise of natural givens, whether or not nature is assumed to be in a state of equilibrium or disequilibrium, energetic balance or overabundance. A periodic expunging of surplus energy does not threaten accumulation and growth; it only reinforces them. Excess remains a reactive category, logically and ontologically dependent on the framework it exceeds.

I would be remiss were I to pass in silence another signature energy dream in political economy, i.e., Smith's ideal of free trade. What this dream envisions is the bubbling of economic energy thanks to a removal of obstacles to its activation, and what it anticipates, in the final instance, is a noncommunist type of synergy.

As Smith sees it, free trade is an instrument for realizing the potential of human sociability and generating wealth for every party involved. A successful pursuit of personal and national enrichment is indicative of the fact that a spiritual-providential purpose has been progressing toward actualization. Grounded on the interdependence of its self-interested, self-loving members, a "civilized society" is one where an individual "stands at all times in need of the co-operation and assistance of great multitudes."[18] Their cooperation is the synergy fueled by the release of heterogeneous energies toward the end of placating individual self-interests. *Free* trade evokes the mutual liberation of self-interested energies. On the international arena, the synergy of self-serving desires spells out peace among freely trading nations, as Ricardo observed in the shadow of Smith: "This pursuit of individual advantage is admirably connected with the universal good of the whole. . . . It diffuses general benefit, and binds together by one common tie of interest and intercourse, the universal society of nations throughout the civilized world."[19]

Peacefully synergetic free trade is a fantasy of the economic energy of rest, with complementary self-interests working together—in the absence of a common plan or blueprint—to facilitate the actualization ("the universal good") of the whole.

In the twentieth century, Milton Friedman took the idea of the free market to its logical conclusion. He concurred with government intervention if, and only if, it served "as a forum for determining the 'rules of the game' and as an umpire to interpret and enforce the rules decided upon."[20] Direct state interferences in the market are, in Friedman's opinion, detrimental because they work against, counteract, and thus restrain economic forces by holding commodities above or below a self-correcting market price.[21] More vividly than previous political economists, Friedman dreams about market economy as a self-regulatory, self-sufficient, self-moving energetic system that takes care of and perfects its operations. This dream elevates the market above its interdependent participants and, consequently, envelops it in an aura of divinity or animality, the supernatural or the natural. Yet, full of and with itself to the brim, the energy of the market relies on the persistence of lack in its participants, whose synergy is only the with-energy of needs resonating with one another. The persistent activation of market dynamics, perversely emulating Aristotle's thought thinking itself, is a putting-to-work predicated on want and, more concretely, hunger.[22]

Whatever the type of economy, it is the sum total of energy synergetically organized, channeled, distributed, and consumed. The energy in question is a mélange of human, machine, animal, vegetal, and physical or elemental forces, with more or less autonomous financial and institutional dynamics thrown into the mix. Owing to the ambiguity of energy's ergon (a work and to work, noun and verb), all the contributors to this sum are both active and passive with respect to its component parts and the whole. Human beings are not only producers, consumers, and subjects of exchange; they are also produced by the very commodities and services they consume, even as their labor is consumed by, or absorbed into, the body of the materials they work on—not to mention the consumption of human resources by capitalist markets and the mysterious-sounding *needs of the economy* to which actual human lives are sacrificed. Everything and everyone can change places at any moment, above all, in economic discourse itself: production is actually the productive consumption of materials, labor, and means of production; consumption is the consumptive production of demand, consumer subjectivities, and value, if we are to believe marginal-utility theorists. With the exception of the elements, the energizing can always become the energized, and vice versa.

Were we to take the promiscuity of economic energy seriously, we would have probably refrained from a destructive-extractive exploitation of natural and human "resources," fueling both productive and consumptive activities. As a matter of fact, we would have recoiled from the very language of resources—their exploitation or, in an ostensibly more benevolent fashion, their management—if we were to accept the thesis that there is no totally passive substratum, which could receive energy's work or enworkment. Finished works also work: rather than an aberration of commodity fetishism, that is the maxim of economo-ontological energy. Along with so-called raw materials, they store and discharge energy, offering resistance to the form imposed upon them or generating value on the superficies of circulation. Synergy is a string of energy's transactions with itself, never identical to itself, swinging between the active-verbal and passive-substantive extremes and swirling across species, kingdoms, orders of beings, animate and inanimate, living matter and long-dead vegetal and animal remnants transformed into natural gas and oil. To economize synergy or to capitalize on it is to let it be good *for* something, conducting it toward life, growth, a quantitative increase of value, etc. that will inevitably revert back to energy. This reversion signals that, regardless of the energetic or synergetic means employed, energy will invariably surface as their end.

Some of the intermediate ends posited in the economic realm will stimulate a future synergy. Others will hamper it, forcing it to fall apart, to fall out with itself. For one, the American dream of *energy independence*—is there any other subtext to the "American dream"?— usually summoned up to garner voter support for fracking and intensified offshore drilling for oil, gives voice to the desire to achieve self-actualization in the absence or at the expense of others, a resistance to synergy that has nothing to do with dependence on oil-producing countries in the Middle East. For another, our economies are accelerating energy's fallout with itself and banking on the breakdown of synergies as they pit workers against each other, against their work (verbal), and their works (substantive); machines against the human labor they replace and thus leave without the means of subsistence; financial growth against the lives of plants, animals, and humans; and the entire destructive-extractive apparatus of production and consumption against the planet. Riven and blocked, alienated from itself, from the synergy that grants it actual existence, energy is corralled in a virtual domain that, expanding like a black hole, wields a deadly force prepared to swallow up everything that is. Ergo, the premonition that, notwithstanding our scrambling to intensify its extraction, energy is dwindling and our world is being swept away along with it.[23]

The paradox of our situation is that the accumulation of energy further blocks and diminishes that which has been accumulated. When it is stored, energy ceases to enwork or emplay itself and so retreats into sheer virtuality, not to be conflated with rest. Financial and industrial capitals are the subsets of general energy accumulation that induce virtualization; the fossils used as sources of fuel are the accumulated reserves of past solar energy. Their distribution is fundamentally unjust, seeing that the energy expended by the many, in a human and nonhuman synergy, is appropriated by the few. Capitalist appropriation spirits away the joint energy—the co-operation and the *opera*—of others and, in the process, divests energy of its actuality, proscribing synergetic relationality. (Overtly or covertly, economic mergers are not cooperative arrangements but hostile takeovers.) But the logic of capital is not the only guilty party here. In addition to capital accumulation, appropriation itself is deeply problematic in that it neglects the synergies of the worker and the work, the elements, machines, and nonhuman living beings, dividing them into proprietors and property to be possessed. Marx's reverie of abolishing private property must be followed through concerning both private ownership and the category of property as such. Imagining the proprietor to be a synergetic ensemble of every economic energy field requires us to recast appropriation in terms of a dispensation of their due back to all the participants in the economy. Communist insistence on *collective ownership* will repeat the mistakes of the past unless what we mean by this well-worn term is *synergetic coexistence*, the energy freely taking a hold of itself as synergy, that is, as an ensemble of all the energies activated, actuated, and actualized together.

Compared to its accumulation, an instantaneous discharge of economic energy is not quite an improvement but rather the other extreme on the continuum of the storage-and-release model. The acceleration of industrial, speculative-financial, and even agricultural cycles along with discoveries in the attendant technoscientific complex permit a more intensive exploitation of human and nonhuman energies. But their sum total cannot grow. The illusion of increase is explicable with regard to the fact that more types of energy are caught up more thoroughly in the nets of capitalist economy. Energies continue to dwindle in the measure in which they are exploited, their synergetic ties severed and enervated.

Trying to hang on to disparate, fugacious, dwindling energies, our economies worry about *energy production*, which "involves converting one form of energy into another form that is needed the most. For example, the sources of chemical energy from fossil fuels and nuclear resources are used to produce approximately 90% of the world's electricity."[24] On the fields of sense cleared by physics, the term is an oxymoron: energy does not come into being and does not drop out of it, but, undergoing endless modulations, simply *is*. Yet, in spite of its nearly unanimous acceptance, this scientific position is too metaphysical. I, myself, would be inclined to entertain the idea of *energy generation*, on the condition that it named a passage from the virtuality of separate energies to their actuality in and as synergy.

Let us assume that *energy production* is a useful concept and take it at face value. Then it will be equivalent to the sum total of economic activity, much wider in scope than the "fuel" on which the work, works, and workers "run." Given that everything and everyone in the world—from stalks of corn to ossified ancient forests, from a plow-dragging buffalo to a factory worker on an assembly line and the machine she operates—registers on the radars of economic rationality solely as a segment in a vast network of resources, the entire system is predicated on the maximum extraction of energy they harbor. In an industrial mode of production, the machine is accorded a special status, considering that it processes energies from the most diverse sources, melding them into a unity (void of synergy): the capitalist industrial product. Capital itself is the difference between the quanta of energy expended in production and those contained in the final product. Whereas physics and classical, precritical political economy tell us that, barring occasional imprecision and aberrations, that difference will be equal to zero, Marxist economic theory declares that the product contains more energy or value than had gone into its manufacture, on account of a portion of the workers', let alone plants' and animals', activity not being compensated. Capital is the production and reproduction of *surplus energy*.

In postindustrial economies, with robust service and hi-tech sectors, the very power-energy-value equation that would have yielded the index of inequality is erased. The worker *is* the work: the two senses of energy's ergon coalesce, without being reconciled, in the same undifferentiated notion. Energy production is the production of the worker, apparently isolated from the activities of plants, animals, material objects, machines, and other workers. Postindustrial economies precipitate a further erosion of synergy, responsible for the retreat of energy from actuality and, by the same stroke, for the virtualization of work. They generate energies already stunted, exhausted, scattered, and consigned to autistic isolation, prevented from finding rest in their consummation or in the sharing of differences. The postindustrial extraction of surplus from human resources is hardly more equitable than its industrial counterpart. If anything, it is more relentless, immediate, and imperceptible than before. With respect to capital, we are all biomass now, powering its growth with our very lives, thoughts, and desires, exchanged for the currency of value in the crux of our being, in clandestine affinity to those plant monocultures that are cultivated as potential biodiesels.

Curiously, the fragmentation of energies in economic activity coincides with their homogenization in the production of fuel spanning combustibles, human food, and livestock feed. A case in point is the corn agribusiness. Much of this grain grown in the U.S. is converted into ethanol at a high environmental cost, from nitrous oxide–releasing fertilizers to coal-powered refineries where it is distilled.[25] Another lion's share of corn, "a crop that alone consumes about one-third of US crop space," is used as feed for livestock.[26] The third major corn-based industry is high fructose corn syrup (HFCS), added to a plethora of American foodstuffs and responsible for the onset of a number of diseases in those who consume it.[27] That is to say, "green" cars and buses, cattle raised for slaughter, and humans all run, in one way or another, on corn. Divided into various resources, they are united by the same substance that animates or activates them, releases their dangerous energies without synergy, but not before inflicting irreparable damage on plant communities and the elements. Whether they are "conventional" fuels or biodiesels, combustibles annihilate the substratum on which they burn and the atmosphere into which their by-products are released. Corn and sugarcane derivatives likewise devastate the human bodies they cram with excess calories in keeping with the needs of capital accumulation. In its current form, energy production is world-destruction; it supplies the fuel for a globalized earth to go up in flames as one.

4

PSYCHOLOGICAL REVERIES

I f energy is at home in economics, then energy *dreams* are on their
native turf in psychology. Dreams are, to a significant extent, wish
fulfillments that give the psyche a taste of the satisfaction it cannot
attain in waking life, blocked by the constraints of the reality principle. What are the wishes that come true in energy dreams, when energy
dreams in us or through us? A craving for indestructibility and the mur-
derous desire to obliterate the entire world, active rest and tireless activity made manifest in perpetual motion . . . It is understandable why these
wishes are self-contradictory. The energy, which could be an object of
desire or its subject, is itself ambiguous, unclassifiable, incongruous with
itself. But wish fulfillment, if only in a dream, is a ramification of energy
that activates every aspect of psychic life, whatever its relation to the
"outside world." Dreaming of energy, we do not hanker after something
absent on a par with other yearned-for objects. Unbeknownst to us, who
are the concretizations of its reveries, energy strives toward itself in and
through us the moment we voice, consciously or not, our desire for it.

The crisis of energy that engulfs our psychic existence is evident in the
ideal of self-control. Whom or what do I control when I control myself?
The mental space-time continuum is a synthetic a priori with energy
configured in synergetic clusters, often at odds with one another. Much
as psychic life can sometimes profit from inner conflict, indicating that,

at least, strife is registered at the level of consciousness, at other times tensions among synergetic bundles can reach a boiling point, forcing them to disintegrate and let go of their energy. Psychosis is a generic label for this disintegration, relinquishing any semblance of self-control and disbanding all synergies into "free" energy bereft of actualization. The debate as to which part of me should hold the reigns and determine my actions thus presupposes the reality of a moderately divided psyche, some of its parts counteracting (undermining the enworkment of) others. At stake in the question of self-control is neither the absolute sovereignty of one faculty over the rest nor a hierarchical sequencing of faculties but an order that ensures the synergy of mental energies. Its cognitive and perceptual corollaries admit attention and concentration, inevitably colored by an affective attachment (yet another type of energy) to that upon which we concentrate. At the same time, psychic synergy is provisional, ready to fall apart at any moment, to turn to another object or to disperse into distraction, a scattering of energies. Only divine attention is constant and utterly concentrated on everything that is, as Plato intimates in *Phaedrus* by painting an indelible image of the soul as a charioteer guiding a pair of "winged horses."

More so than Aristotle, whose psychological method could be called an *archaeology of the soul* with the deeper strata of vegetal and animal vitalities overlaid by human reason, Plato is alive to the concurrent activation and counteraction of diverse psychic energies. The Platonic soul is comprised of the "joint potentialities" of the charioteer and the horses: ἐοικέτω δὴ συμφύτῳ δυνάμει ὑποπτέρου ζεύγους τε καὶ ἡνιόχου (*Phaedrus* 246a). *Sumphutō dunamei*, an expression I translated as *joint potentialities*, literally says "the potencies that grow together" or "the potencies that partake of the same nature." Only in the gods, Plato explains, is this with-growth of diverse kinds of dunamis perfectly balanced and good, which is why pure divine energy moves in a flawless circle, strives to itself alone, and finds the good in itself. Finite living beings are, for their part, doomed to look for the good outside themselves, connected as they are to an external energy outlet. Worse yet, in human souls, one of the winged horses is of a noble lineage and the other is unruly (246a–b). "Therefore, in our case the driving is necessarily difficult and troublesome [χαλεπὴ δὴ καὶ δύσκολος ἐξ ἀνάγκης ἡ περὶ ἡμᾶς ἡνιόχησις]" (246b). As antagonistic potencies pull the human soul in opposite directions, its energy wanes and the transition to the full actuality of energeia is foreclosed. Not so for divine souls, where, thanks to the with-growth of their parts, dunamis is one and the same with a synergetically consolidated energeia.

Plato's allegorization of psychic energies introduces a vertical axis of movement when a finite ensouled being comes on the scene. This being is pulled down toward the earth by a horse that, having lost its wings, partakes of evil—ὁ τῆς κάκης ἵππος μετέχων—and upward toward the ether of ideas by its winged, celestial double (247b). On the one hand, Plotinus and the Gnostics will relate the heaviness of the soul's defective potency to the corporeality of the body, its material character being the evil that weighs us down. In what amounts to a gross oversimplification of Plato, the more enmeshed are the two, the greater the gravitational pull toward the earth. With Aristotle, on the other hand, the body itself will be empowered and disempowered with respect to energy. Curtly put, 1. a body shorn of a soul is energy-less and harbors a potentiality to receive an energetic charge from its animating principle; yet 2. the energy of the soul is that of a natural body that lives up to its good and, ontologically, to what it is supposed to be over and above life's dormant potentiality. The synergy of the soul and the body is the body working in concert with itself across the temporal gap separating its potential being from actual existence.

To hone the sense of psychosomatic synergy, I propose that we read, as closely as possible, one of Aristotle's definitions of the soul in *De anima*: "Hence, the soul is the first actuality [*entelecheia*] of a natural body having life in potentiality [διὸ ἡ ψυχή ἐστιν ἐντελέχεια ἡ πρώτη σώματος φυσικοῦ δυνάμει ζωὴν ἔχονοτς]" (412a, 27–28). On par with energeia, *entelechy* is normally rendered in English as "actuality," given that in the Aristotelian vernacular both concepts are contrasted to the potentiality of dunamis. It designates the specific work of the soul on the body, delivered to its end (telos) and allowed to be all it can be. This activation, this psychic activity, completes the body, perfecting it, and guiding its potentialities toward fulfillment. Before its enendment or ensoulment, a natural body is free of energy, which is not the same as saying that it is without power. Indeed, it is traversed by various powers, yet none of them lets the body stand out from the flux of physical forces as a bearer of a definite shape or form—the end as the instantiation of energy's substantive dimension. If the soul is the first actuality, enendment, enworkment, or energy of a body primed for life, then, prior to its intervention, the body is not yet actual; it is destitute of energy, at the mercy of forces totally external to it, thrown any which way independently of its proper end.

Still, the soul is no more imposed onto the body in Aristotle than form is thrust onto matter from the outside. The entities that surround us in everyday life are hylomorphic units, in which form (*morphé* or *eidos*) belongs together with matter (*hulé*). Living bodies are ensouled (that is, they have a unique form) and given coherence not by virtue of some extraneous ends but through the cultivation of a vitality appropriate to them. The *en-* of *entelechy*, similar to the *en-* of *energy*, teases the telos out of the entity to which it belongs, if potentially. Entelechy is the specific way in which energy puts itself to work in a living, or a potentially living, body. In comparison with ergon, its telos lacks the ambiguity of a process *or* a product, to work *or* a work; it is the end, that toward which a living being strives. Nonetheless, the end of entelechy is not final. The soul is the *first* (*prōté*) enendment of the body, with further teloi to follow from the enworkment of its organs (412b, 5–6). Naively understood, it is the initial, rudimentary energy, which is potential vis-à-vis the acts of living that will ensue, as well as the good itself that is the final end of all things. But the first in Aristotle is more than that. Take, for instance, first philosophy, self-sufficient and complete in its foundational capacity. Or the first enworkment of the body by the soul, the enworkment that is not deficient but plentiful and energetically robust relative to those that will follow. Or, again, the idea that the initial actualization is already at the end, in view of the potentialities it has actualized. The energy of the end is the beginning and it is placed at the beginning in Aristotle's initial attempt to demarcate the territory of the soul.

It is telling, in fact, that the definition of the psyche draws a circle— the end of entelechy comes first—that will endure as the soul's privileged spatial representation. This circularity does not necessarily emulate Platonic divinities, perfect and perfectly self-sufficient; it could well be the effect of a body's self-relation, its first actuation, underlying all other activities, and hence its first actuality that is the soul. The body is to the soul as the world is to the unmoved mover: the second constituent in each pair is the outcome of a self-relation in the first, to take a little further Losev's theses we have commented upon in the opening chapter. A body that pertains to the realm of *phusis* energizes itself, pulls itself up by its bootstraps, latches onto itself, and, through this first act, lays the ground for subsequent activity. Its entelechy enlivens it by lending it coherence, gathering it together with itself, if only precariously and impermanently. Irreducible to a receptacle, a site where some work foreign to it takes place, the dispersed body is the means to an end of self-activation, that is, a work in progress and a forever provisional outcome.

The first energy of the body that is the soul is not the same as the power ascribed to dunamei-potentialities, the precursors of entelechy. That the soul is powerless is made plain in its signature receptivity to nourishment in its vegetal, to sensations in its animal, and to knowledge in its human instantiations. Powers are claimed by the body, said to be a "dynamic thing"—τὸ δὲ σῶμα τὸ δυνάμει ὄν (413a, 1–3)—but in the psyche they are quelled by the minimal energetic accomplishment of potentiality. This rest is, in turn, disturbed by the accumulation of disparate kinds of vitality within the same soul, from the vegetal to the animal and human, the earlier strata forming the deepest psychic substratum.

In *Nicomachean Ethics* Aristotle admits the soul's cleavage into two parts: one that possesses logos and the other deprived of it: δύ᾽ εἶναι μέρη τῆς ψυχῆς, τό τε λόγον ἔχον καὶ τὸ ἄλογον (1102a, 29–30). In its rendezvous with the *alogon*—a psychic compartment destitute of logos, usually assumed to derive from nonhuman living beings and persisting in the human soul—ancient psycho-logy, which is a discourse (logos) of and about the psyche, reaches its limit, even as it opens exciting venues for the study of synergy at the level of mental faculties and states. Before all else, this encounter demonstrates that human entelechy cannot come to pass alone in isolation from vegetal and animal vitalities *in us*. The enworkment or enendment of the human soul is a collaborative travail, which, intriguingly, relies on modes of life that may come into conflict and negate the psychic form we habitually avouch as our own.

Alongside the alogical part which cannot be reasoned with, there is in the human psyche another cross-section that, though similarly devoid of logos, participates in the activity of reason: ἔοικε δὲ καὶ ἄλλη τις φύσις τῆς ψυχῆς ἄλογος εἶναι, μετέχουσα μέντοι ᾗ λόγου (1102b, 14–15). A person is self-restrained, in control of her- or himself, provided that the alogical faculty makes good on its participation in logos and is heedful to discourse and reasoning. Participation, *methexis*, exemplifies how mental synergy works without canceling out the differences among the participants. In lieu of an anthropic privilege, logos itself is participatory and synergetic beyond the human in the human, as it draws the one and the other (or others) into a conversation, which need not rely on words and in which each maintains her, his, or its discreteness. The energy of logos articulates, assembles the expressing and the expressed, together with the parties to a dialogue, even when it is only I who discourse with myself in the intimacy of the psyche or when one or more of the parties is nonhuman, inside and outside the human. Its energy, then, is erotic in the sense with which Freud will imbue eros in consonance with the thinkers of antiquity, as the psychic drive toward aggregation. And, vice versa, moving away from the principle of methexis, we do not slide down a hierarchy of living beings, among which plants are probably the most participative of all, given that humans and animals are alive by dint of taking part in their nutritive soul.[1] Only when the energy of logos fails to pass into a synergy does the propensity for sharing diminish, leading to an autistic enclosure in oneself and a crisis in the subject and its world. Reason's monologue is the tombstone on the grave of psychic energy.

―――― ∞ ――――

The problems of ancient psychology I have raised here are still our problems, not because nothing has changed in human subject-formation over the period that has elapsed between Plato or Aristotle and us, but because, like them, we have no means of enunciating psychological insights other than by and through the psyche. The subject matter of psychology is inseparable from its object of study, which is the subject: the working and the work proceed in tandem, completing an energetic circle. Indirectly, in his "economic" theory of the drives and in the elaboration of conscious and unconscious topographies, Freud, too, salutes Greek thought for the unrivaled attention it has paid to both modalities of psychic energy. Psycho-analysis is an exceptional legatee eager to retrace the outlines of psychic schisms in its very method, an inheritor of ancient psycho-logy who transcribes its logos into the language of analysis, palatable to post-Kantian philosophy. Out of fidelity to the Delphic injunction *Know thyself!* that condenses the entire energy of Greek psychology, Freud makes self-analysis a cornerstone of his clinical procedure and theoretical investigations. In this, his method radically deviates from behaviorism and, more so still, from the statistical techniques determining the contemporary shape of the discipline that prefer to forget about the existence of a psyche and to replace it with actual behavioral patterns or their average distributions in a population. A psychology without a psyche and without logos, without the discourse of the psyche by and on itself (I hesitate to write "a psychology without a soul," resorting to a phrase that could be perversely understood as a nostalgic tribute to an old theological and metaphysical category), ours is a psychology whose energy has been sapped. In the absence of its object of study, which is also its subject, let alone a method that would lead toward or from—from-toward—the psyche, psychology functions nowadays without working and without work, so much so that it forecloses the potentialities of self-knowing. It is for this reason that I turn to psychoanalysis for a reinvigorated approach to (and of) psychic energy.

The Freudian psyche is divisible and divided, and its study needs to be *analytical* to remain faithful to that irreducible fissuring. This axiom applies not just to the major partitions of psychic topography—the system *Cs./Uncs.* or its later mutation into the tripartite structure of id-ego-superego—and not just to the inner differentiation of libidinal energy, distributed according to the intensities of its attachment to an object (cathexis), but also to "primary" processes, notably infantile narcissism. Rebuffing the assumption that at the early stages of its development the psyche (that is to say, psychic energy) is undifferentiated, Freud humbly accepts the inadequacies of analytic devices in discriminating between fine-grained libidinal clusters: "Finally, as regards the differentiation of psychical energies [*die Unterscheidung der psychischen Energien*], we are led to the conclusion that to begin with, during the state of narcissism, they exist together and that they are undifferentiated for our coarse analysis [*und für unsere grobe Analyse ununterscheidbar sind*]; not until there is object-cathexis is it possible to discriminate a sexual energy—the libido—from an energy of the ego-instincts" (*SE* 14:71).[2] Analytical tools cannot peer below a certain threshold of events in the psyche. Although, in the preobjective state of primary narcissism, libidinal energy is minutely differentiated, we can spot it no sooner than it attaches itself to an object. And we can, likewise, discern impressions deriving from the external world only after they have passed the perceptual threshold and become vivid enough to be registered by our senses. Below the point of discernibility, sexual libidinal energy appears to be one with that of the ego-instincts, turned to itself as the other. To avoid erroneous conjectures, such as "What I cannot see or hear does not exist" or "Prior to ego-formation, there is no psychic differentiation," Freud is predisposed to viewing this energetic constellation as a coexistence of discrete energies in closer proximity (*zunächst*) to one another than a psychoanalyst is capable of discerning—a proximity that suggests a synergy of libidinal and ego-instinctual drives.

Freudian psychoanalysis equates the intensity of psychic energy to the strength of cathexis and of the corresponding power of repression. Unrepressed unconscious contents exhibit "a small amount of energy," but "as soon as the basically obnoxious idea exceeds a certain degree of strength, the conflict becomes an actual one [*wird der Konflikt aktuell*], and it is precisely this activation [*Aktivierung*] that leads to repression" (*SE* 14:152). "So that," Freud continues, "where repression is concerned, an increase of energic cathexis [*Energiebesetzung*] operates in the same sense as an approach to the unconscious" (*SE* 14:152). Psychic energy is thus a quantum of strength, which correlatively determines the strength of repression. Which means that Freud's inability to dissociate energy from force puts him on a collision course with ancient psychology. Despite picturing the psyche as a conglomerate of "unbound," uncathected, or free-floating and the more or less strongly cathected or "bound" energies, his method is psycho*dynamic*, emphasizing, in a quintessentially modern gesture, psychological dunamis—potency or potentiality. (The science of physics, similarly, colludes with the anti-Aristotelian notion of energy, which it treats, and mistreats, by recourse to the laws of thermo*dynamics*.) Regardless, his insistence on the actuality and activation of conflict and repression displays Freud's commitment to the ancient conception of energeia as an actuation or activation of what has been previously hidden in potentia.

The birth of the subject is, according to psychoanalysis, conditioned by the transformations of psychic energy that, at times, at the height of repression, turns against itself, working against the enworkment of other mental processes. Freud subscribes to Spinozan monism and to the laws of physics when he represents the gamut between affects and ideas as a single energetic scale, its vicissitudes depending on the intensities of repression at any given point. Ideas are the representations of instincts that could, just as well, "vanish from the conscious" or be "held back from consciousness" under the influence of repression (*SE* 14:153). In them, for the time being, energy announces itself, dreams itself in the transparent cobelonging of the representing (working) and the represented (work). The conscious is energy's self-consciousness.

When they vanish from consciousness, ideas can disappear "without a trace," or they can be subject to a "*transformation* into *affects*, and especially into *anxiety*, of the psychical energies of *instincts* [die *Umsetzung* der psychischen Energien der *Triebe* in *Affekte* und ganz besonders in *Angst*]" (*SE* 14:153). As ideas disappear "without a trace," energy recoils to the deepest strata of the unconscious, whence it produces other, more distorted symptoms of repressed desire. Short of fading away along with ideas, psychic energy becomes affect: here Freud gives a nod of approval to Spinoza's hunch that feelings and emotions are but muddled, unclarified ideas. What, to Spinoza, was substance, modified into a deluge of manifestations, is, for Freud, "the psychic energies of the drive," now as ideational representations of instincts conscious of themselves, now as the mutability of affect, now as virtually traceless, almost inaccessible, unconscious blockages. It bears repeating though that Spinozan substance is one in contrast to a multitude of its epiphenomenal appearances; Freudian psychic energies, for their part, are a priori differentiated, as plural as the objects they are bound to, or more numerous still, if we count their finer, preobjective varieties. The primacy of difference in psychic life is yet another piece of evidence for the nonmetaphysical quality of psychoanalytic energy.

Jung concurs, in an essay on "The Type-Problem in Poetry," that "like physical energy, libido passes through every conceivable transformation; we find ample evidence of this in the phantasies of the unconscious and in myths."[3] He also interjects fantasies between the ideational and the affective expressions of psychic energy, neither as transparent as the former nor as befuddled as the latter: half-dreaming, half-awake. Fantasies, like ideas, are libidinal self-representations, and they are, similar to affects, encrypted in a symbolic form: "These phantasies are primarily self-representations of the energetic transformation processes, which follow their natural and established laws, their determined 'way' ['Weg'] of evolution. This way signifies both the line or curve of the optimum energetic discharge as well as the corresponding result in work [Arbeitsleistung]."[4] Jung flounders, nevertheless, in his attempt to specify the nature of "natural and established laws" governing libidinal energy. What forces its torsions, transformations, and encryptions? Which resistances does it encounter and how does it circumvent them? What are the circumstances, where we experience vacillations between ideas, fantasies, and affects? "Libido," Jung explains, "as an energy concept is a quantitative formula for the phenomena of life, which are certainly of varying intensity."[5] Couched in terms of quantitative energy differentials, the explanation is deficient, if we concede that it leaves outside its theoretical frame both qualitative changes in libidinal self-manifestations and differences in cathexis that cannot be squared with the varying intensities in the "phenomena of life." On this point Freud is more comprehensive and more nuanced than Jung.

Where Freud's promising colleague, who would later become his rival, excels is in the recovery of the ancient notion of energy for psychology, combined with Vedic thought. The "way" recalls Rita, "the right path" of the Vedas, as well as the two halves of energy itself: the working "line or curve of the optimum energetic discharge" and the work that results from it, *Arbeitsleistung* (the language of work, *Arbeit*, is, I believe, deliberate and mindful of energy's etymology). It is, of course, the libido that moves along the path and *is* this very path, whose symbolic self-representation in fantasy is a decidedly Jungian energy dream.

With the view to preserving the dynamism of the way, and so relating to energy more as a working process than a thing or a product, Jung forewarns his readers against hypostatizing the libido, made "synonymous with psychic energy." The latter is not to be treated as "a psychic *force*," *nicht eine psychische* Kraft, available for hypostasis, but as "a concept denoting intensities or values," *als einen Begriff für Intensitäten oder Werte.*[6] Insulated from "force," the libido is a good match for ancient energeia, whose actuality precludes the power of potentialities. At the same time, Jung overreacts against the attributions of a real identity to the libido; the prohibition on its hypostatization risks overshadowing the substantive work-dimension of energy, leaving the phenomenon in its integrity and the possibility of its quiescence in tatters. For the Aristotelian "energy of rest," we should look elsewhere in psychoanalytic literature, namely the Freudian discovery of what lies beyond the pleasure principle.

———— ❦ ————

The overall goal of Freud's psychoanalytic endeavor was to usher in, within the ambit of "terminable analysis," a fragile psychic synergy conducive to mental health. Nevertheless, an abiding acumen of his work, summed up in the notion of "interminable analysis," cannot help but interfere with this goal: try as we may, libidinal energy is going to undermine itself, to part against itself, to put itself to work or into play for two or more contradictory reasons, with incompatible objectives. Very often, the unconscious subverts the work of consciousness, as in the case of negation, in which what is meant is the inverse of what is said. When I assert "This is not my mother!" about a woman seen in a dream, I unconsciously convey that it *is* my mother, for otherwise there would be no sense in denying her symbolic appearance. The nontotalizable multiplicity of libidinal energies evinces mutual incompatibility among some of its strands.

That the self-undercutting of psychic energy is not accidental, that it is explicable with reference to a generic pattern of libidinal currents and countercurrents, was corroborated in Freud's discovery of the death drive in his 1920 *Beyond the Pleasure Principle*. As he writes with a hindsight of nearly twenty years in *Civilization and Its Discontents*, "I drew the conclusion [in *Beyond the Pleasure Principle*] that, besides the instinct to preserve living substance and to join it into ever larger units, there must exist another, contrary instinct seeking to dissolve those units and to bring them back to their primaeval, inorganic state. That is to say, as well as Eros there was an instinct of death. The phenomena of life could be explained from the concurrent or mutually opposed action of these two instincts" (*SE* 21:119). An implacable analysis, the dissolution of complex unities by the death drive is put to work simultaneously with the propensity to build up these very unities. What is more, "concurrent and mutually opposed action" proceeds from the same libido, the same energy that, in activating, deactivates itself, and vice versa. And all this transpires, despite tireless invocations of death, on the side of life, whose phenomena are explicable through the halting rhythm—the uneven, self-contradictory enworkment of construction and destruction.

In Newtonian physics every action presupposes an opposed and equal reaction; in Freudian psychoanalysis most actions happen *concurrently* with the reaction, which is usually unequal to the quanta of libidinal energy the original action binds or releases. Psychic life is the movement of attraction and repulsion, assembly and segregation, of social units—from a couple and an oedipal family to a state—as much as of the subject itself. Adamant about the inclusion of counterwork in the workings of the libido, Freud writes in another text that "only by the joint and mutually opposed action of the two primal instincts—Eros and the death-instinct—, never by one or the other alone, can we explain the rich multiplicity of the phenomena of life [*nur das Zusammen- und Gegeneinanderwirken beider Urtriebe Eros und Todestrieb erklärt die Buntheit der Lebenserscheinungen, niemals einer von ihnen allein*]" (*SE* 23:243).

Zusammenwirken, the collaboration or the synergy of Eros and Thanatos, *always* goes hand in hand with their *Gegeneinanderwirken*, working against one another. Try as we may, we cannot forgo tension, contention, and conflict in mental synergy that inherently strives to undo that which it creates. Freud, in effect, avows the togetherness and apartness, the collaboration and the counterwork, of the principles of togetherness and apartness: Eros and Thanatos, synthesis and analysis. Before the discovery of "beyond the pleasure principle," psycho*analysis* was skewed in its focus on the principle of aggregation, Eros, to which it had assimilated libidinal energy in its entirely. But was *its* analytic drive not, already then, an (unconscious) intimation of the death drive?

Adding to the difficulties of trying to capture a panoramic picture of Freudian psycho-energetics is the fact that the pleasure principle imitates death, with its desaturation of tensions, and the death drive feeds on the pleasure of repetition. Freud commences his *Beyond the Pleasure Principle* with the decision "to relate pleasure and unpleasure to the quantity of excitation that is present in the mind but is not in any way 'bound' [*mit der Quantität der im Seelenleben vorhandenen – und nicht irgendwie gebundenen –*]; and to relate them in such a manner that unpleasure corresponds to an *increase* in the quantity of excitation and pleasure to a *diminution*" (*SE* 18:7–8). The "quantity of excitation" refers to psychic energy, either bound in object-cathexes or unbound and yet undetermined. Pleasure, for Freud, denotes a decrease in the free-floating form of energy that passes into a state of being bound to an object, wherein it finds respite. Accordingly, the pleasure principle resonates with the energy of rest, psychoanalytically articulated in the aim of converting "freely mobile cathectic energy into a mainly quiescent (tonic) cathexis [*frei bewegliche Besetzungsenergie in vorwiegend ruhende (tonische) Besetzung umzuwandeln*]" (*SE* 18:62). United with its object, the drive is fulfilled and the *work!* of energy is enriched with the substantive facet mitigating its verbal signification, which reaches us as an inexorable imperative. Unbound from the object, "freely mobile" energy is working without a work, provoking unpleasure, and causing an increase in the quantity of excitation. The preeminent function of object-cathexis is, therefore, to fasten the two sides of ergon to one another.

Death erupts, in this scheme of things, as the harshest instantiation of the reality principle, a most consequential interference in the self-enclosed circuit of psychic energy that cuts the bond between the object and the ego. The task of mourning—significantly treated as a *work*, *Trauerarbeit*—is to rebind the energy previously cathected to the lost object to another object or to a memory of the one that has vanished, adjusting the mourner to its "real" unavailability. Melancholia responds to this task poorly, as the melancholic stubbornly clings to the lost object, continuing to pour inordinate amounts of energy into cathecting it, or gives free reign to negative, liberated energy previously associated with the object. It follows that melancholia disallows the quiescence of the drive in cathexis and liberates the working without a work that countervails the tendency of the psychic apparatus to switch energy from the unbound into the bound variety. Freud documents this energetic anomaly with the utmost precision, "The complex of melancholia behaves like an open wound, drawing to itself cathectic energies [*zieht von allen Seiten Besetzungsenergien an sich*]—which in the transference neuroses we have called 'anticathexes'—from all directions, and emptying the ego until it is totally impoverished" (*SE* 14:253).

A hallmark quality of "freely mobile" (*frei beweglichen*) energies is that they "press towards discharge" at a time when primary psychic processes are arrested in their development and have not yet had a chance to mature—through the inhibition of excitation in cathexis—to the level of secondary, conscious processes (*SE* 18:34–35). What is experienced as a flurry of intense activity is only an acting-out, a misleading form of energy deprived of quiescence or rest in a cathected object. The subject's incapacity to bind energy into mental bundles, deferring its discharge, would, according to Freud, "provoke a disturbance analogous to a traumatic neurosis; and only after the binding has been accomplished would it be possible for the dominance of the pleasure principle [*die Herrschaft des Lustprinzips*] (and of its modification, the reality principle) to proceed unhindered" (*SE* 18:35). In its mutilated form, psychic energy that acts out independently of the pleasure or reality principles is the compulsion to repeat, *Wiederholungszwang*, which is the shortest route toward discharging massive quanta of instinct. It generates acts outside the range of actuality, oblivious to the reality principle and standing in the way of the pleasure principle. Analytic work—working through (*Be-*, *Ver-*, *Auf-*, or *Ausarbeitung*) rather than acting out (*Ausagieren*)—is, conversely, a mindful repetition of what occasioned the excitation of the drive in the first place, such that the energy associated with it would now be bound, becoming amenable to an alteration into the stuff of secondary, conscious processes. The work of analysis is this labor of cathecting freely floating energy, though the true object of its cathexes is often substituted with the psychoanalyst her- or himself, giving rise to transference and counter-transference. By its mere practice, psychoanalysis throws light on a fissure in energy between acting and working, an immediate externalization of what is put in play in psychic life and an expression of subjective interiority with the mediation of the pleasure and reality principles.

If unbound, freely mobile energy finds itself on the side of the death drive rather than the pleasure principle, that is because it is propitious to the dissolution of synthetic unities, beginning with object-cathexes. Its dynamism is that of deactualization, a virtualizing power of pure and abstract potentiality. The death drive carries the kind of energy that, in working, unworks itself, undoes the works of previous bonds forged in psychic life. As a result, there are two main psychic energies that could never coexist in synergy with one another: "life instincts and death instincts." So wide is the chasm between them that Freud is willing to qualify his psychoanalysis *tout court* as "dualistic," *dualistische*, in contrast to Jung's "monistic" libido theory: *Jungs Libidotheorie ist dagegen eine monistische ... (SE* 18:53). At times organized into oppositional clusters, such originary difference, if I may call it that, invalidates the claim that Freud adopts a thermo-psycho-dynamic model of energy whereby the mental apparatus is a mechanical system. For, what machine throws wrenches into its own wheels?

⤙⤚

The machine image of the psyche is prima facie irresistible whenever one deals with a conception of energy that oversteps the perimeter of its purely physical field of operation and covers the psyche. In the Cartesian universe, extended-corporeal substance was figured as a machine, a met-aphor, from which the indivisible substance of the mind was exempt. It would seem logical that, once we acknowledge the mind's divisibility, as psychoanalysts do, the laws that used to govern body-machines would apply to the mental apparatus as well. That is the position of Jacques Lacan, who adverts to how "in Freud something is talked about, which is not talked about in Hegel, namely energy" and how "energy . . . is a notion which can only emerge once there are machines [*L'énergie . . . est une notion qui ne peut apparaître qu'à partir du moment où il ya des machines*]."[7]

Lacan's statement is wrong for two reasons. First, Hegel talks of nothing but energy in the movement of Spirit towards the fullness of actuality, its Wirklichkeit reconciling in itself the rational and the real. Second, the notion of energy is independent of the world of machines, as the Aristotelian coinage suggests. Its quiescence, which in a mechanis-tically determined universe stands for the termination of functioning, is the fulfillment of potentialities or the binding of impulses in object-cathexes. It is true that a crucial apothegm of psychoanalysis states that the psyche is immanently divisible, but this feature does not have to replicate machine disassembly into component parts. Should we look for analogies, divisions in the life of the mind would approximate a fur-rowed surface of the earth or, in the context of a "dualistic" conception of energy, the splitting of an atom pitted against itself.

Mechanistic explanations would have been adequate if psychic life hinged on but one impulse, the compulsion to repeat. Acting out, the sub-ject acts like a machine, mindlessly (more exactly: unconsciously) engag-ing in the same behavior over and over again. Lacan himself pays tribute to the "insistence" of repetition compulsion, "translated into French as *automatisme de repetition*" (61), and insinuates that its operations are those of an automaton. Such explanations may even hold for the offset-ting tendency toward homeostasis in the pleasure principle, but they can-not accommodate the concurrence of psychic action and counteraction. Monistic perspectives on energy are totalizing and, in being so, uphold the Lacanian dictum that "energetics is also a metaphysics [*l'énergétique est aussi une métaphysique*]" (61). Freudian "dualism," however, affirms the originary difference in and between psychic energies. *Beyond the Pleasure Principle* broaches the prospects of energy beyond energetics and of a psychology unburdened of metaphysical trappings.

Freud and Lacan agree that the transposition of biological, let alone thermodynamic, realities onto the psyche will not advance psychological investigations by one iota. Behaviorism is the main, albeit unnamed, culprit here: all it occupies itself with are behavioral inputs and outputs, mimicking the materials and by-products of metabolism. "There are quantities of energy the organism assimilates, by various means, and there's what, taking everything into account—muscular expenditure, effort, dejections—comes out of the mechanism. To be sure, the laws of thermodynamics are respected—there is degradation of energy [*il y a degradation de l'énergie*]. But about everything that happens inside, we know absolutely nothing" (96–97). Putting to one side Lacan's misguided attribution of the "degradation of energy" to the theory of thermodynamics, his argument holds: psychology cannot afford to reduce the psyche to a black box, a deep structure in the form of a question mark, where the inputs are *somehow* deposited, altered, and meted out again. Laudable psychological theories and practices should aim to repair the tie not between the energy of the surface and that of the depth but between the heterogeneous energetic levels of psychic life. In Freudian psychoanalysis the symptom extends one of the bridges between these levels without canceling out their heterogeneity; in Lacanian thought the conduit for interactions between them is the contrast between the symbolic, the imaginary, and the real, revisited on the heels of his critique leveled at the physicist-biologist reductionism.

Be this as it may, I do not think that the most interesting side of "depth psychology" is what it stumbles upon in or inventively projects onto the deep recesses of the psyche. The fascinating bit is how the body's essentially superficial self-relation—say, the psychosomatic interface constitutive of the Lacanian real—enters the fray and invests "depth" with meaning. Disconnected from the surface, the energy of psychological depth is the home turf of traditional metaphysics, and it teeters dangerously on the precipice of theology with its conception of the soul as the withdrawn realm of interiority, where faith demands a thoroughgoing cathexis of pent-up energy to objects, to an object, or to *the* object that is God. Far from a panacea, the *mechanical* energy of the psychological surface befits the posthumous incarnation of metaphysics in "energetics" and describes the behaviorist construal of human and animal comportment. Psychoanalytically examined, this mutilated energy is responsible for repetition compulsion, the hurried acting-out that leaves no time for "mental digestion." Behaviorism as a whole indulges in an epistemological repetition compulsion and unconsciously projects its deficient method onto its subjects. Wearing the mantle of scientificity, it is the ersatz of Nietzsche's incomparably more stimulating proposal to dissolve the doer in the deed, to make all energy anonymous, active, as well as outwardly accessible, and to turn the substantive being individuated in an actor into an afterthought and a fiction.

Freud bemoans, in *Moses and Monotheism*, the slow progress of psychoanalysis on the track of understanding the nexus between disparate energetic levels, a deficiency that results in "our complete ignorance of the *dynamic* nature of the mental processes [*unserer völligen Unwissenheit über die* dynamische *Natur der seelischen Vorgänge*]" (*SE* 23:97). "We tell ourselves," he continues, "that what distinguishes a conscious idea from a preconscious one, and the latter from the unconscious one, can only be a modification, or perhaps a different distribution, of psychical energy [*kann nichts anderes sein als eine Modifikation, vielleicht auch eine andere Verteilung der psychischen Energie*]. We talk of cathexes and hypercathexes, but beyond this we are without any knowledge on the subject or even any starting-point for a serviceable working hypothesis" (*SE* 23:97). A study of the modifications and variations in the distribution of psychic energy leaves psychoanalysts in the dark, lest they manage to trace these surface phenomena to libidinal production. Lacan's concern with our insensitivity to "everything that happens inside" echoes this damning verdict penned by Freud. And, besides, the modifications of libidinal energy are out of touch with the subterranean counterwork of the death drive. Not only is the segregation of cathexes into objects too coarse to elucidate differentiations and transitions from primary to secondary psychic processes, but also the very notion of cathexis is too crude to capture the dynamism of psychic life.

Most commentators are inclined to impute this self-berating ret-
rospective glance to the embittered mood of an exiled, old, and dying
author. In my assessment, nevertheless, the questioning of psychic
energy in the late 1930s, that is to say in the last years of Freud's life, stayed
faithful to the undefinability of the concept (if a concept it is), which
had troubled Aristotle. Some of the strongest evidence for my argument
may be gleaned from *An Outline of Psychoanalysis,* drafted in 1938 and
published posthumously in 1940. As Freud intimates there, "it will be
entirely in accordance with our expectations if the basic concepts and
principles of the new science (instinct, nervous energy [*nervöse Energie*], etc.) remain for a considerable time no less indeterminate [*unbes-
timmt bleiben*] than those of the older sciences (force, mass, attraction,
etc.)" (*SE* 23:159). The indeterminacy of the basic principle of "nervous
energy" is there to stay: as in Aristotle, it derives from the impossibility
of defining and the necessity of comprehending this energy by analogy
between psychic forces and their physical counterparts. But the mechanism of analogy, too, falters here. So indefinite is the subject matter
that Freud characterizes it as an "undisclosed secret," *nicht enthüllten
Geheimnis*: "Here we have approached the still undisclosed secret of the
nature of the psychical. We assume, as other natural sciences have led
us to expect, that in mental life some kind of energy is at work [*eine Art
von Energie tätig ist*]; but we have nothing to go upon which will enable
us to come nearer to a knowledge of it by analogies with other forms
of energy" (*SE* 23:163–64). The energy at work, the enworkment of psychic life, cannot, on balance, be compared to the thermal, mechanical,
atomic, or other sorts of energy. Freud's words, "some kind of," *eine Art
von,* stand at the apex of vagueness, which cannot help but affect the
entire analytic enterprise, given that such vagueness strikes at the heart
of what this enterprise tries to decipher—"the nature of the psychical."

How unfair, in light of these nonaccidental hesitations, it is to lump
Freudian theory together with energetics, as Lacan does! Instead of taking into account Freud's radical self-critique, which suspends the possibility of a straightforward analogy between psychic and other kinds
of energy, the French psychoanalyst represents his predecessor's jagged
intellectual itinerary as so many leaps from one energy system to another.

The early Freud, he submits, "tried to build a theory of the functioning of the nervous system by showing that the brain operates as a buffer-organ between man and reality, as a homeostatic organ. And he then comes up against, he stumbles on, the dream world. He realizes that the brain is a dream machine [*une machine à rêver*]" (76). That other energy dream—presumably less conservative both in form and content—which extracts psychic energy precisely from dreams, does not depart from the machine trope of the brain. It merely switches the modes of the apparatus's functioning. The Lacanian Freud is a machine operator of psychoanalysis who sometimes reinvents the contraption he is working on without ever interrogating its mechanical makeup. Re-embedding the mind in the body, he allegedly expounds the workings of both based on the same physio-logic principles. Which means that his findings can be effortlessly assimilated by the cognitive sciences that similarly picture mental processes as algorithms spewed by the brain's computing machine.

Take, for instance, Freud's affirmation in *The Interpretation of Dreams* that "dream-thoughts are entirely rational and are constructed with an expenditure of all the psychical energy of which we are capable. They have their place among thought-processes that have not become conscious—processes from which, after some modification, our conscious thoughts, too, arise" (*SE* 5:506). Lacan could conceivably argue that the rationality of dream-thoughts inserts them into a chain of mental inputs, processing, and outputs. But a mechanical system cannot account for the overdetermination, condensation, and displacement of libidinal energy into these image-thoughts. Calling the brain a *dream machine* adds nothing to our appreciation of how psychic life works, what is enworked or emplayed in it, and why. Nor does it explain how conscious thoughts arise "after some modification" from those not yet conscious. Heavy with meaning and sense, the beginning of thinking is the energy dream of the dreams themselves and, in other ways, of the dreamers and psychoanalytic interpreters. Thinking is, emphatically, not information processing.

Indeed, the stage of "processing" conceals more than it reveals, inasmuch as it gives us to understand that the transition from inputs to outputs is smooth and automatic. It hides everything Freudian psychoanalysis concentrates on: the inner tensions, resistance, splitting, negations, partial cathexes of psychic energy, and the complex, self-contradictory inscription—the registration *and* unawareness—of the "inputs." No matter how intelligent, machines cannot leave binary logic behind: they either work or do not work. They cannot, at any rate, work in not working or not work in working, which is the condition of enworkment (energy) in general and of psychic enworkment in particular. Furthermore, in a machine, energy flows are distinct from products or outputs, whereas in psychic life, and preeminently in dreams, conflicted, self-contradictory energy is the means and the end, the activity and its outcome, just as it was for Aristotle. The "secret . . . nature of the psychical" touches upon this exceptional proximity between the whole and a part, energeia and its psychological instantiation, where the process and the product—the two dimensions of ergon—coincide.

When it comes to Lacan's own system of thought, our problematic is encoded in every invocation of "acts," from *passage à l'acte* to *signifying act*. Taking this observation as our point of departure, we note that the categories of the imaginary, the symbolic, and the real are indexed to variations in the production, distribution, and actualization of psychic energy. The real, which "is without fissure," *le réel est sans fissure* (98), is the nonsignifying acting-out of unconscious desire and is roughly equivalent to unbound libidinal quantities in Freud, who, for his part, insinuates that what presents itself as an unfissured facade to the analyst contains fine-grained differences inaccessible to the analytical tools at our disposal. It is psychic energy detained in pure potentiality, full of force yet powerless to devolve some of its physical power to the symbolic. Acting-out is *not* (yet) acting. If an act is to take place, *passage à l'acte* must shake off the temptation to act out.

It may so happen, of course, that the real is "without fissure" only in itself, for at this stage there is not yet a self-relation (i.e., the minimal synergy) required for the enworkment of energy, both as the means and the end of psychic life. The act of signification will abstain from introducing a previously nonexistent gap into the real. It will only indicate and perhaps exaggerate what has been always already there, a potentiality actualized: "For the furrows opened up by the signifier in the real world will certainly seek out the gaps—in order to widen them—that the real world as an entity [*étant*] offers the signifier."[8] The symbolic places a mirror before the real, reflecting previously unnoticed inner differentiations, marking or re-marking the gap that has endured there, and walking potentiality toward its actualization in the act of signification. Those who speak Lacanese know that *signifying act* is a tautology, since, in and of itself, "an act is signifying [*l'acte est signifiant*]."[9] But this tautology is also rewarding to the degree that it articulates the discombobulated halves of psychic energy, the working and the work. Nothing less than the subject is enworked, activated, put into play in an act of signification, which is the act *proper*, "a true act," *un acte véritable*.[10]

Where does this tautology leave the Lacanian real? Unexpectedly, we are led to a deduction that the real is not the actual so long as it lacks the symbolic supports that reiterate and, in reiterating, return it to itself. And, to some extent, it will never be actual, given the gaps that cut it off from the possibility of a symbolic repetition, as well as from itself in the absence of a self-relation. In the real, dunamis is stuck, its *passage à l'acte* barred, much like in the state of trauma it is frequently identified with. Reality itself is equivocal, seeing that it encompasses "the two terms, *Wirklichkeit* and *Realität*, that Freud distinguishes . . . the second being especially reserved for psychic reality."[11] Psychic *Realität* is bereft of actuality (hence of energy) not because it is entirely ideal, imagined, or lives on solely in our heads but because it has not (yet) actualized itself by handing itself over to the act, which is *per definitionem* signifying. The powerlessness of this act will be at variance with the real; it will come about as a result of an excess of psychic energy in a subject who is the product and the process of signification, a subject who, by and as a sign, convokes itself into being, in which it is nothing and to which it brings nothing but a reiteration of the gaps that have peppered potentiality.

And that is as close as Lacan gets to the Aristotelian energy of rest. The French psychoanalyst locates the birth of the subject from a self-generated act within the region "beyond the pleasure principle," which, for him, means beyond energy. To the best of his ability, he tries to maintain

Wirklichkeit at arm's length from Realität, the former connoting "phenomena of energy and of nature," and the latter—those of psychic life (ethics, *la Chose, das Ding*). What Freud took as an inner splintering of psychic energy into the bound and the unbound, Lacan deems to be an abyss between energetics and the (energy-less?) symbolic realm: "The phenomena of energy and of nature always tend in the direction of an equalization of levels of difference [*une égalisation de dénivellation*]. In the order of the message and of the calculation of chances, to the extent that information increases, the difference in levels becomes more differentiated. . . . Everything we call language can be organized around this basic principle."[12] But what if, by symmetrically aligning the two principles, Lacan replaces the metaphysics of energetics with the metaphysics of difference? Raised to a *principle*, everything (including difference and differentiation, which, in a vast majority of cases, is the branching out of the One) will supply the grist for a metaphysical mill. How is the metaphysical question of energy different from the metaphysics of difference? So long as there is energy, its forms and quanta are not all the same, notwithstanding the entropic tendency toward equalization or equilibrium.[13]

Between the real and the symbolic, we cannot forget the imaginary, a concept with a long and convoluted history in Lacan's oeuvre. The prime focus of the imaginary order is the relation of the ego and the specular image at the mirror stage, when the previously fragmented sense of my body is integrated into a unity vis-à-vis the other who faces me and in whom I recognize myself. I must first separate from myself to find myself as an external self-representation, making "alienation . . . constitutive of the imaginary order. Alienation is the imaginary as such."[14] Imbibing the negative charge of symbolic differentiation, which Lacan subtracts from the ontology of energy on the grounds that the latter is too natural or naturalistic, the imaginary is powered by the energy of dissociation, analysis, and holding-apart.[15] Its end result is, therefore, a certain make-believe, the pretense of an order without fissure, bearing uncanny resemblance to the unmarked gap of the real and yet hinging on an ongoing, incomplete overcoming of constitutive alienation. Ultimately, the discernment of a difference between I and myself, or between myself and my reflection, sends me a memento mori. Seeing himself in the mirror, as both the same as and other to himself, the "human animal is capable of imagining himself mortal—which does not mean that he would not do so without his symbiosis with the symbolic, but rather that, without the gap that alienates him from his own image, this symbiosis with the symbolic, in which he constitutes himself as subject to death, could not have occurred."[16]

Worth noting is the fact that the imaginary is activated, almost in its entirety, by the energy of the surface, fueled by specular images at "the threshold of the visible world."[17] The unmarked void of the real is the deep and inexhaustible source of the subject, symbolically and superfluously-superficially born from the signifying act that re-marks, and in re-marking widens, that void. The imaginary, in turn, dawns in a "dual relation," in a separation between and an imperfect overlap of two surfaces—the seeing and the seen. Although this order is not hallucinatory, as Lacan often underscores, it is more fantastic and phantasmic than the symbolic. Its energy is dreamy and captivating, enthralling and seductive. The energy of the symbolic is an act ideally free of potentiality, a signifying enworkment, whose product is the subject itself. That of the imaginary is the dynamic—the dunamis—of becoming-oneself in a back and forth of specular alienation and identification. Nowhere else but in the symbolic will the imaginary come to itself outside, or beside, itself and be itself by receiving its actuality from the signifying act. ("The imaginary experience is inscribed in the register of the symbolic as early on as you can think it."[18])

To sum up: for all the theoretical wrangling, the priority Aristotle accorded to actuality over potentiality still holds sway, probably thanks to Hegel's mediation, in Lacan's thought. The signifying act is of the highest order, as the end of psychic development and the means for conveying the image or remarking the unmarked gap of the real. Beyond energetics, the energy of the symbolic is energeia pure and simple.

Attention, among other cognitive functions, fleshes out the modulations of energy in psychoanalysis. If Freud recommends that the analyst assume the position of listening with free-floating attention to the free

associations of the analysand, then we can surmise that the analytic situation begins, and constantly renews itself, by tapping into the unbound
psychic energies of both. But the consonance of two (or more) distracted
attentions is only the beginning. The success of psychoanalysis, notably
of its interpretative *work*, will be contingent upon how well the energetic flows of the attending and that which is attended to are cathected,
bound into object bundles, or, in a word of Lacan, symbolized. When
they complain that psychoanalytic formations—chief among them, the
oedipal triangle—arrest desire, Gilles Deleuze and Félix Guattari foil the
transition from free association to interpretation and imply that attention should be concurrent with unbound psychic energy, unrestrained
by object-cathexes, and, in this hyperbolic format, faithful to the streams
of desire itself. Much like Henri Bergson, the coauthors of *Anti-Oedipus*
prefer the active-verbal aspects of energy's work to those that are pro-
visionally accomplished and substantive, or substantialized. On their
watch, desire attains no rest, however temporary, and attention is internally supplanted—qua hyperattention—by perpetual distraction.

As an alternative to the metaphysical fixation on pure, amaranthine
attention and to the maximalist postmetaphysical celebration of the
inverse, "dynamic" shuttling of the mind, I suggest taking a closer look
at the finite capacity to attend to the world—to the surface, the skin of
things—as the innermost expression of psychic energy. Granted, the
moment an object falls into the focus of my attention, some quantities
of the libido are tied to it and are detained together with it in a fixedly
attentive regard. My interest in anything whatsoever is in sync with the
sudden protrusion of something that captivates me from the general and
mostly homogeneous perceptual background. That is the affective infrastructure of fundamental phenomenological correlations: the intending
and the intended, the seeing and the seen, the feeling and the felt, etc.
Lasting as it may be, the bond of cathexis is reconfigured ad nauseam,
as attention, together with the quanta of energy it involves, allowing for
shifts to new experiential objects. A possible exception to this rule is
the state of meditation, wherein the attentive attitude achieves a certain
measure of rest. Even so, the fight against distraction—a disposition,
which, on the positive side, turns us toward other, previously unnoticed

stimuli—can have an adverse effect on psychic life by initiating the pet-
rifaction associated with Medusa's deadening gaze.

Without reproducing the disastrous effects of metaphysics, the psychoanalysis of attention discriminates between the energy of the surface
and that of depth in this comportment. Psychic stratification into conscious and unconscious systems shows that explicit attentiveness to a
stimulus in one's environment is, more often than not, a symptom for
concerns and even fixations not readily discernible, including from the
vantage point of the attentive subjects themselves. Giving one's time,
binding one's energy and one's senses to X may be a manifestation of a
subterranean attention to Y, which cannot be aired out in the open. At
that deeper level of energetic investment in a forbidden object of desire,
cathexis can get knotted and entrenched, leading to a condition whereby
the drive is stuck (fixated). The fixity of an attentive regard can, at any
instant, flip into fixation, the arresting of energy in its purely substantive
state. Alternatively, psychic counterwork, in forcing us to pay attention
to something, may do so in order to distract us from something else—
death, tabooed desires, and so on. Retrieving and nurturing the patients'
energy of depth, the therapeutic relation neither extracts anything from
nor destroys their psyches. In an analytic interaction, the analysand's
psychic life is reconstructed, becoming less knotted and freer without,
by the same token, falling into the trap of unbound libidinal flows.

We could say in this regard, relying on Lacan's vocabulary, that we are
never able to access the real (in the guise of the totally undifferentiated)
without the imaginary and symbolic prisms preinterpreting it. In the act
of paying attention, the object reaches me leavened with significance. It
is, thus, an *act* in the Lacanian sense of signification, be it in the midst
of the not yet explicitly symbolized perception. Attention to an object is
inscribed in the register of the symbolic as early on as you can perceive
it. But to recognize yourself, the attending subject, in a preoccupation
with an object, you have to endure radical self-alienation, interrupting
the predictable, because preorganized, flow of psychic energy from one
thing to another to another . . . not in order to mark the real itself, but to
remark the demarcation etched on it. To stop acting compulsively, or acting out, and to attend to attention itself in a subjectivating *passage à l'acte.*

Could this pause send us back to the residues of the energy of rest in the ever-shifting modulations of "dynamic" attention and desire?

There seems to be, for all that, no respite both in everyday cognitive or perceptual experience and in its psychoanalytic iteration. Pressing needs and novel stimulations instigate restlessness, hyperactivity, and the plague of attention deficit disorder. The binding of energy in object-cathexes is by no means a remedy, not the least because its exaggerated *stasis* may be a symptom of fixation and of the drive's painful paralysis. That is where, I think, psychoanalytic theories ought to be complemented with the phenomenological investigations of how consciousness works.

Although desire and the dynamism of psychic life, which it condenses and represents, do not know a final resolution, Edmund Husserl holds throughout his writings that anything experienced and thought is in and of itself the fulfillment of the experiencing and the thinking. The very fabric of our mental existence is woven of energies, in which the intending belongs together with the intended, the seeing with the seen, the hearing with the heard, the judging with the judged . . . The fulfillment of "empty" intentionality in intuition is the synergy of mental workings with that on which—or with that toward which—they work. The attending gaze, ear, or hand *rests* in the attended-to pen, sound of a door being shut, or fur being stroked. Abounding in the energy of rest whenever my intentions accord with the evidence received in intuition, such fulfillment is a positively Aristotelian residue in Husserl's philosophy. Admittedly, though, the quelling of intentionality in intuition, of mental dunamis in energeia, is irrelevant to the limit cases of the insatiate desire for the other or the thought of death. There is no evidence for these intentional comportments, no matter how attentive we are. The other and death, exerting tremendous influence on my existence, are nowhere to be found in my world; they enwork or emplay my psychic life on behalf of another energy, which towers over the power of attraction binding empty intentions and fulfilled intuitions.

The combined psychoanalytic-phenomenological doctrine of attention is germane to the elucidation of psychic energy in other areas of cognition and perception. With one proviso: the attentive attitude is both a segment of our mental existence and its animating factor. Attention puts the psyche to work or into play—whether at the level of consciousness, by singling out and placing objects in its crisply focused field, or at the level of the unconscious, by attending to the world outside the sphere of our awareness, where hyperattention borders on distraction. If the conscious modality of attention is the moment of gathering, binding, concentrating, then the unconscious modality, which has a lot in common with distraction, is that of dispersion, unbinding, dissipating. From its commencement, the enworkment of the psyche feverishly begins working against itself. Pure conscious attention is (can only be) the mark of divinity, driven ever upward by two winged horses, following Plato's narrative, or monotonously working in harmony with itself: a deus "qua" machina. In the earthlings that we are, internally conflicted and pulled in every which direction at once, it is but a utopian offshoot of our energy dreams, the fantasy of unconditional self-gathering. Contrary to canonical authors, from Plato through St. Augustine to Simone Weil, I do not see reasons for despair in this predicament. Care for the psyche, its cultivation if you will, beseeches us to care for the unconscious—learning to live and to work with its counterwork. Assenting in this way to finitude, non-self-transparency, and inner contradictoriness is the sine qua non of our training in how to draw energy from the imperfect synergy of our psychic acts.

5

POLITICAL FANTASIES

As I first noted in "Opening Words," *Energy Dreams* is a sequel to my 2015 book *Pyropolitics: When the World Is Ablaze.* The core of my own energy dream, which instigated the composition of that text, was the desire to rescue the political from the dead weights of institutionalization, codification, formalization, bureaucratization, and technocratic neutralization, while sidestepping the metaphysical trap of relentless activity. To address this challenge, I had to turn to the enigmatic, non- or premetaphysical, thinking of the elements—primarily, the earth and fire—in ancient Greece, as well as to non-Western, Indian and Chinese, elemental thought. Later on, I grew cognizant of the fact that my efforts were indebted more to dunamis than to energeia: political energy boiled in the uncontrollable capacity of the flame to effect transformations, induce revolutionary upheavals, motivate, illuminate meaningful ideas and impart the heat of action or . . . burn to the ground whatever or whomever "caught" fire. In fire and the earth I saw the Janus-faced manifestations of *stasis*, of revolt and quiescence. A flaming principle was gradually buried under the illusory stability of the terrestrial sphere in a sequence of events as much political and as they were semantic, seeing that *static* and *status quo* acquired the sense of immobility and unchangeability.

I felt I had no choice but to portray the postmetaphysical era as that of smoldering cinders and ashes, the fire going out, energy exhausted, nonviolence achieved at the expense of waning existential vibrancy. To be honest, I still find this narrative persuasive, even if I also take it to be inadequate without a thinking of energeia that blends the two meanings of *stasis* in what I have been calling, with Vladimir Bibikhin, *the energy of rest*. Is there a way in which political activity and organization, the working and the work, might obey the protocols of this Aristotelian not-quite-concept?

Should we dare to dream with Aristotle and Kant, we would recognize that the politics of energeia is allergic both to a permanent revolution and to institutional proceduralism. It is, strangely, a politics of powerlessness, so long as we interpret this last word positively, and not in terms of a divestment or lack of power. Accordingly, a salient example of energeia in political thought is Kant's perpetual peace, which the German philosopher prudently distances from "the peace of the cemeteries" at the outset of his essay. That is his most cherished energy dream, reveling in the energetic rest of humanity that has carried its capacity to reason to its completion, left behind the irrational waging of war, and regulated public affairs on every plane, from the local to the cosmopolitan, on transcendental grounds. Or, more precisely, it is the "sweet dream," *süßen Traum*,[1] of all philosophers, of philosophy itself.

Kant believes that his dream is not an empty, threadbare musing and that his insight into the energetically approaching actuality of perpetual peace sets him apart from other thinkers who long for it. "Not just an empty idea," *keine leere Idee*, perpetual peace is a matter of self-actualization (literally, "making itself actual") "in an infinite process of gradual approximation [*Unendliche fortschreitenden Annäherung wirklich zu machen*]."[2] Becoming-actual is setting-to-work, energizing. What is set to work here is perpetual peace, which, as perpetual, is always at work, before and after it is concretely instituted. The energy Kant singles out and nourishes his philosophical discourse with is one that transforms theory into practice, a void idea into a task to be discharged. In other words, and contrary to the habitual take on the transcendental gap between the regulative Idea and reality, perpetual peace is the vanishing difference between a mere idea and a task, the difference that discharges enormous quanta of energy for the movement of history, politics, and human development. This cleft cannot be altogether sealed; the "infinite approximation" of the one to the other, of working for genuine peace to the work of peace itself, inaugurates a dialectics without synthesis, more open and sophisticated than Hegel's. The gap lies not between political energy and an unattainable ideal that motivates it, but within energy itself, separating its verbal from its substantive expressions.

Kant's perpetual peace is the highest end of human history that actualizes itself and, in so doing, salvages the Aristotelian telos that, in the last analysis, is the good, *tō agathon*. Unique among "every art and every investigation, and likewise every practical pursuit or undertaking that seems to aim at some good" (*Nic. Eth.* 1094a), the art of the political strives toward a common, most universal good, prompting Aristotle to crown political science as the queen of the sciences. The aspiration to perpetual peace goes a step further: it dares to envision the *accomplishment* of the common good, its enworkment and actuality, its energeia prescribed by self-legislating reason and, indeed, inseparable from reason's authority. It reverts to what Aristotle aims at with his notion of happiness, *eudaimonia*, now distributed among all humanity. To the degree that reason's authority activates the principles of coexistence, to which any rational being subscribes, up to the cosmopolitan sharing of the earth, the need to resort to power, physical force, violence, or external coercion is obviated. A powerless politics par excellence, perpetual peace is a vivid snapshot of the energy of rest in modern political philosophy.

With a penchant for cynicism, postmodernity urges us to wake up from this dream, which implies, in some sense, waking up from philosophy as such. Whose reason gains an upper hand under the guise of universality? Which peace does it institute? What energy brings it about? In whose interest is it sustained? These are, assuredly, pertinent questions, though they also exhibit something of an overreaction to universality, to a good common to two or more people. Instead of the good, there are only partly overlapping interests; instead of instrumental goals, intermediate means for furthering them. In tandem with these substitutions, political energy fades, and capitulates to economic forces, is diverted into a fatigue-inducing reaction to events, or, worse, to a reaction to a reaction. Having proudly emancipated ourselves from a much-maligned classical teleology, we have thrown the baby out with the bath water, that is, have gotten rid of the positive notion of *end* as though it were an authoritarian fiction. No sooner than we rouse from the dream of perpetual peace—the open-ended end that reactivated Aristotle's energeia in Kantian thought—we fall into a worse nightmare of endless means that circumscribe the political field as we know it.

Contemplating energy, we normally associate it with the beginning: the cosmic Big Bang; the initial push a deistic God gave to the world in Creation before recoiling from what has been created; the moment of ignition, the spark that activates a previously dormant engine . . . In twentieth-century political thought, Hannah Arendt stresses the energizing influence of beginning ever anew. Elliptically, silently, *energy* signals an activation of the beginning, which sets itself to work or puts itself into play in everything that will follow. Watery, fiery, or, at times, a puzzling elemental mix of the two, it is what flows or flames up from a source hidden in the past and, immediately upon being activated, finds itself on the cusp of being extinguished, drained away, and spent. So potent is this preconception that it colors our understanding and experience of time, itself coterminous with the initial explosion, whose aftermath it accompanies until the total diminution and exhaustion of the force released at the origin. It goes without saying that the end is negatively related to energy, as the stage at which the energetic charge ebbs away without remainder, becoming absent.

But what of the energy of the end—the end that energizes, that imbues with energy, say, by pulling something or someone toward itself and bestowing sense on that which or the one whom it attracts? What if the quanta of energy, no longer translatable into quantitative terms, actually increased at the end? For the Greeks, after all, enworkment is enendment, *entelecheia*. Telos is that end which animates beings in the flux of becoming, draws them toward itself *from within*, by virtue of impersonal discernment, summoning each singularly, according to what or who it is. "Yet another fiction, another metaphysical daydream," you will frown as you brush off the outdated arsenal of teleology. And you will be in a good company with Heidegger and Derrida, who, as we have found out, group energeia together with other prominent keywords in the glossary of metaphysics. It remains to be seen, however, whether the cultural and ideological hegemony of means is really an improvement over the ancient teloi it has supplanted. Everywhere there is frantic activity, plenty of energy expended, though, as we have seen, without an end and to no end. The end vanishes, even as the discourses and concerns around it grow exponentially, provoking fear, excitement, and vague anticipation of the final exhaustion of being and energy, of being *as* energy, or of being *through* human addiction to particular types of energy, above all, fossil fuels.

Today, *the end* can broadly denote only two things: 1. a realizable, attainable objective—in the Heideggerian idiom, the ready-to-hand and 2. cessation, the not-present-at-hand. Objectives are never final; they are intermediate milestones, the means for new provisional "ends" in the infinite regress of instrumental rationality. Cessation, in its turn, is linked to an attenuated conception of finality. It brings to naught what was unfolding before the end grasped as a standstill, an arbitrary interruption, foreign to the consummation of movement and the satisfaction of rest. In each case, the end signifies a relative or absolute desaturation of energy on the treadmill of instrumental reasoning and in the cut of termination without a clear term. The poverty of its meaning is due to our imputation to it of the sense of a limit, a razor-sharp edge where a spatial surface or a temporal line abruptly drops, rather than a boundary or a border, for instance, between motley worlds.

Since we are already speaking Kant's language of borders and limits, *Grenzen* and *Schranken*, reminiscent of Aristotle's telos and peras, we would do well to note that there are also ethical ends, which obviously do not fit within, and in fact resist, the logic of means.[3] Where do these ends derive from? Do we not give them to ourselves, or receive them as unquestionable givens, as part of a philosophical compensation package for the tyranny of instrumentality? A noncalculative notion of the end, reminiscent of the thought of antiquity, is a fetishistic substitute for the lost object, which is not this or that thing but the entire teleological paradigm. Crucial to its enunciation, Kantian ethics (and, to a lesser extent, his aesthetics) performs the work of mourning over the corpse of ancient teleology, aspiring, along the way, to conjugate the end with another energy, to put it to work otherwise. A world of programmatically established and technically carried out objectives and abrupt terminations would be unlivable, obliging us to carve out small niches for ends without means in the midst of the reality of means without end. So understood, the premonitions and anxieties about the end of the world strike us from the same structural place on the margins of instrumental rationality as ethical care for the ends-in-themselves.

Heidegger's thesis regarding being-toward-death, too, participates in the lineage of fetishistic substitutes for the techno-metaphysical constellation of ending, the substitutes that momentarily jolt the entropic inertia of our ideology but do not really trammel it. With all fairness, the author of *Being and Time* appreciates the manifold ontological energies of the end, be they 1. teleological, 2. categorial-objective, or 3. existential. So, 1. a ripening fruit, in reaching its end, fulfills itself; 2. the rain that stops is no longer present-at-hand and the bread that is used up is no more ready-to-hand; 3. Dasein ends in unfulfillment and, at the same time, *is* already its end, a being-toward-the-end, *Sein zum Ende*.[4] The finite energy of human existence, oriented toward and defined by its noncalculable end, is but a parenthesis between the teleology of the living, kept intact in Heidegger as much as in Kant, and the physical availability or unavailability of things. Dasein is put to work or into play—thrown and projected: *thrown projection* is another term for existential energy— vis-à-vis its unsurpassable, unique, and individuating end, a lonely exception within the overwhelming orders shaped by biological, physical, and technical forces.

Let us see, next, how the three types of ending bear upon the world. By and large, the end of the world is thought of, dreamed about, or dreaded as if it followed the course of ending proper to something *in* the world. We imagine the world's abrupt termination, often for political or geopolitical reasons, a point in time when it ceases to be present-at-hand or ready-to-hand, disappears, melts away like a cloud that finished shedding rain or is used up like bread that has been devoured. Only this conception of the end, furtively tied to the second law of thermodynamics, assumes the entropic fizzling out of being and the abatement of energy generally ascribed to nihilism. Wholly embedded within the networks of means and ends, it does not begin to atone for the overextension of instrumentality into every province of existence. The question it assiduously ignores is: When God is out of the picture, to whose hand is the world that has just ended no longer present?

According to theological interpretations, the end of the world is akin to the harvesting of a ripe fruit and all preceding history is analogous to its maturation. The Book of Revelation is exquisitely instructive in this respect. As angels come down to earth, some of them armed with sickles, a call resounds: "Put in your sharp sickle and gather the clusters from the vine of the earth, because her grapes are ripe [ἤκμασαν]" (14:18). In the allegory, the ripe grapes are the sinners, thrown into the "great winepress of God's wrath" (14:19), from which the righteous are saved. Eschatologically, the world ends in the fulfillment of its divinely ordained destiny and of prophecy, a global fruit rounded off and prepared, through its gradual coming to fruition, to instigate a new beginning, denuding the seed it has contained. Unless it ferments in continued existence after history (or History) has tapered off. The energy of the end actualizes, activates, and animates the first beginning in circling back to a more perfect order than that of the fallen world. One discovers here a kind of spiritual instrumentalization that converts the end of the world into a means for rising to a higher ontological plane, and it could well be that the treatment of existence here-below as a vehicle for spirit supplied the blueprint for our techno-metaphysics committed to the unbridled and smugly spiritless manipulation of matter.

If, *per contra*, our ears are attuned to the existential tonality of this syntagma, then the end of the world is—neither cessation nor fulfillment—the world's very being as being-toward-the-end. Claiming that the world is finite is not enough. It *is* finitude. There is no such thing as the world *as such*, immune to existential, hermeneutical, or perspectival considerations. The world is the end; so long as it exists, it is ending. The end of the world is the end of the end, a formulation that is as tautological as it is suggestive, in that it accentuates the end by way of its doubling *and* brings it to a close. Perhaps radical indecision alone, hovering between affirmation and negation, can reclaim "the end" from the teleological hierarchies of old and from the role of a fetishistic supplement to instrumentality assigned to it by Kant.

Given the existential equation of the world and the end, the energy of the world is the energy of the end, the setting to work or the putting into play of its worldhood, that is, finite time. It would be erroneous to think that the world's chronologically coherent timeline comes to an abrupt halt at the hour of the apocalypse, measured since 1947 on the "Doomsday Clock" by members of the Science and Security Board at Chicago's *Bulletin of the Atomic Scientists*. (In 2016 the clock showed 23:57—three minutes to the end of the world.) The world takes its place and time (in a word: happens) in the interval, in the in-between where nothing happens, where there is no dissonance between activity and rest, and, therefore, where Aristotelian energeia, the energy of rest on the hither side of production and relaxation, rules. Similarly, within the frames of reference constructed by Judeo-Christian theology, the world unfolds in the interval separating Creation from Judgment Day, in the difference between this world (in Hebrew: *ha-ʿolam ha-zeh*) and the world to come (*ha-ʿolam ha-ba*), or, by divine grace, between the first and the second comings of Christ. What is the end of the interval, the interval qua end? Total instrumentality alone is blind to the world's end as a constitutive, inherent, sense-giving border, and explicates it otherwise, as a prohibitive limit, as the interruption of what had been before that limit was reached.

Still, to my taste, none of the above is satisfactory, because we remain straitjacketed by the precise grammar of "*the* end of *the* world," *la fin du monde*,[5] uncongenial to phenomenology and to existentialism, notably to their adage that the world is not one. The one world is a theological, metaphysical, political, and economic fable that dons the cloak of self-evident reality vouchsafed by the one God, or, more recently, by the one financial idol, Capital, instigating the official movement of globalization (*mondialisation*, worldization). Not by chance, the motto of the 2008 Chinese Olympic Games was "One World, One Dream," to which I am tempted to add "One Energy, One End." For, as I put it elsewhere: "When all is One, all is lost; the One *is*, but may also not be, easily succumbing to negation in a single stroke. The One is more vulnerable and fragile than the many."[6]

Our cultures and governments, in the East and in the West, asseverate this unity and are obsessed with its impending demise. Their endgame is this: they put the end of the world to work for the sake of perpetuating the unsustainable existence of the affluent few at the expense of the impoverished many and the planet itself. Through an ideological sleight of hand, availing themselves of a dirty synecdoche, those on the profiting side of this dream represent *their* privileged world as *the* world and pretend that no other kind exists nor is possible. Endangering *their* world becomes tantamount to harming *the* world as a whole.

Affectively binding us to the thought, more unconscious than conscious, that we must chose between the status quo and the abyss, the champions of the current global "order" play the part of the *katechon*, the restrainer, the one who defers the future of no future, traditionally correlated with the coming of the Antichrist, as in 2 Thessalonians 2:6–7: "And now you know what is holding him [the lawless one, the Antichrist] back, so that he may be revealed at the proper time. For the secret power of lawlessness is already at work [ενεργειται]; but the one who now holds it back will continue to do so till he is taken out of the way." Against the energy of the end, against the lawlessness (*anomia*) secretly, sotto voce, at work in the world, the imperial restrainer exercises the sovereign power Carl Schmitt will later accentuate in *The* Nomos *of the Earth*.[7] Thoroughly reactive, the counterwork of the katechon opposes the opposition to law that, irreducible to a collection of legal statutes, is the world's regularized arrangement, the ensemble of order and orientation (the nomos) lending the whole its coherence at the price of many other valid existential interpretations. This counterwork, or anti-energy, is rampant in contemporary politics: with the locus of sovereignty displaced indefinitely, political action is but a reaction on every side of the "friend-enemy divide." And the end of the world is, probably, the most reaction-inducing vision of all.

Those committed to theories of foundationalism will associate the focus on the end and the desire to delay its arrival with the draining of energy from the beginning, the weakening or, conversely, the ossification of the principles that dynamically steer a given unity of order and orientation. Compelling as this idea sounds, it falls prey to the bias of the beginning, of energy's emanation from an original source. It does not inquire into the meaning of *the energy of the end*, where the end is the work (ergon) and the world working itself out in the absence of unified origins, foundations, or immutable principles. The energy of the end as the work or the enworkment of time: temporalizing.

More than that, the political entropy hypothesis is insensitive to the ends (of the worlds) that persist beneath the self-unfolding beginning. Which world's end are we afraid of? The first? The second? The third? The second world, or the Eastern bloc of Communist countries, ended in 1989; the third is ineluctably postapocalyptic, forced to survive in a state of outright worldlessness. World history does not break off when the world ends, for the very reason that it is a history of the never-ending ends of worlds, an unrelenting catastrophe that, in Walter Benjamin's memorable pronouncement, befalls the victims of "civilization" and "progress." (Isn't this disenchanted insight into the—historical—essence of history chillingly similar to the protracted, intractable, unending end of metaphysics?) Instead of putting the foreboding of an apocalypse to work for the purpose of consolidating the political ontology of domination, another energy of the end, seething in Benjamin's texts, gives voice to the ends of the world that have happened and keep happening as we speak, so as to entitle the privileged few *to have a world*.

If we turn the tables, or invert perspectives, then *the* end of *the* world would come to signify a beginning of worlds. The beginning is not a better, higher, more secure unified origin, not an *arché*, and not a sound basis for another hierarchy; it is an anarchy of worlds-ends. The energy of ends, indissociable from that of worlds, is anarchic. It enworks or emplays permeable borders in place of limits and opens intervals in space-time wherein worlds come about, are eventuated. To go to my world's end (i.e., to the end of the end or to the worldhood—the being and the time—of the world) in the thick of this indomitable multiplicity is the exact opposite of being isolated, as on a deserted island. In and at the end, I will chance upon the in-between that articulates my world with many others, not necessarily of a human variety. So imagined, an end of a world is the place of possible encounters, of borders that lightly graze or barely touch one another in the new beginning.

How is the end—notably, of the world—related to possibility? Let me recall, *en passant*, Aristotle's fierce insistence in *Metaphysics*, that energeia is not-dunamis, a statement to which he accords the status of a quasi-definition. Extricated from the capacities to bring to a close what is yet incomplete, energeia outshines power; the energy of the end is the end's powerlessness. It is in this context that Aristotle invents a name for a teleology-crowning fullness, where rest is melded with movement and actuality with activity.[8] A *finite* energy of ends, for its part, precludes atemporal presence and lack alike. Alien to it is the teleological standard of accomplishment, on the one hand, and the craving for—the will to—power that is enduringly frustrated insofar as we implicitly evaluate it with regard to that standard, on the other. We comprehend existential possibilities that, in Heidegger's text, stand "higher than actuality," *höher als die Wirklichkeit*,[9] in terms of dunamis uncoupled from energeia. Yet, the enworkment or the emplayment of worlds or ends combines possibility with actuality differently, on the hither side of the present-at-hand and the ready-to-hand. It concerns the actuality of possibility as possibility, the existential actuality of the world as a plenitude foreign to the dialectics of accomplishment and incompletion. Violence, motivating the imposition of extraneous ends and the endless torsions of mere potentiality, is neutralized not apolitically but peacefully, where one end of a world approximates and at times touches another.

I have discussed, somewhat tersely, the theme—the end of the world—and an anarchic variation on the theme—an end of a world. Their energies part ways largely as a result of the exclusionary effects of limits on the former and the unruly openness of borders on the latter. There are, however, additional options for modifying this syntagma. I have in mind *the end of a world* and *an end of the world*. The end of a world is death, which, for Heidegger at least, is impregnated with the energy that trumps all other ends (i.e., pragmatic objectives) and is unexhausted in the cessation of biological life. Finitude sets uniqueness beyond individuality to work by giving us a shared foretaste of an unsharable fate, of the end that, though inevitable, cannot be anticipated in the beginning and that doles out to each world the energy peculiar to it. Across this anarchy of ends, it activates an anarchic community, a destiny without a coherent destination, a "powerless superior power," *die ohnmächtige . . . Übermacht*,[10] virtually indistinguishable from energeia.

This is, finally, what *an end of the world* conveys: a thousand deaths, ends, times, terms, terminations, borders, or edges awaiting our common plane of existence, the earthly infrastructure fatefully caught up in the worldly mesh of meaning and sense. The many ends of the same world hint at a plurality of means through which life could be destroyed, a variety of detours, as Freud says apropos of the death drive, leading to the same outcome: global climate change that would make the planet inhospitable to humans and other species, nuclear annihilation, or what have you. But the countless ends awaiting the world are more than winding routes toward the abyss. Were we to stop here, at the means of coming to an end, we would have acceded to the tyranny of pure instrumentality.

An end of the world also says something about life that, at the intersection of world and earth, receives its energy from multiple sources, its material conditions of possibility dispersed among the elements: fresh air, fertile soil, and certain living beings, including vitamin-synthesizing bacteria and plants. By no means a totality, more than one in the one, life, along with a litany of meanings indexed to it, can nonetheless be extinguished as soon as any of the ingredients that put it into play is taken out of the equation. Granted that the result is formally analogous to doing away in a single stroke with the One that has absorbed the All, there is still a tremendous difference between these two events. The world can run into its end from various directions because everything that supports and sustains it—not transcendentally but from within, by participating in it—is singular and irreplaceable, unlike sections of a totality.

In lieu of a coherent principle and a secure origin, the energy that holds it together is a synergy, a precarious togetherness that cannot afford to surrender but a morsel from what, jointly, forges the world and puts life to work or into play. The moment it lets go of singular plurality, the world falls apart, comes undone at the seams, and meets one of its ends, as numerous as the beginnings.

———— ∞∞∞ ————

That the end is not tantamount to cessation is a critical piece of the puzzle in the political panorama of our philosophy of energy. Now, the divergence of cessation from political ends is borne out in Georges Sorel's *Reflections on Violence*. Revolutionary energy peaks during a proletarian general strike, which is more than just a hindrance to productive activity governed by capitalism. The uncompromising stance of general strikers pivots on the decision to disengage from economic, social, and natural necessities, violating every law, from the laws of supply and demand for commodified labor to the laws of nature, including the relentless force of hunger as a motivation to work and the struggle for individual survival. This decision is intensely political, despite the economic provenance of the initial disruption itself. It wins the energy of class struggle back not by plunging into action or activism but by putting an end to work that reproduces the workers' physical existence as an afterthought to the production and reproduction of capital.

Despite his insistence on the proletariat's disengagement from the rules of the capitalist game, later on adopted by the adherents of the *Autonomia* movement in Italy, Sorel highlights the interdependence of clashing political energies, as well as of their entropic tendencies. Calling a socialist government, which waters class struggle down, *the dictatorship of incapacity*, Sorel writes: "Before the working class could also accept this *dictatorship of incapacity*, it must itself become as stupid as the bourgeoisie and must lose all revolutionary energy [*énergie révolutionnaire*], whilst at the same time its masters would have lost all capitalist energy [*énergie capitaliste*]."[11] The dictatorship of incapacity, rooted in a halfhearted compromise, is the exact opposite of the strikers' energetic powerlessness, anchored in their rejection of the productive capacity bestowed upon them by capital. The dictatorship of incapacity perpetuates an oppressive political-economic system; the refusal to work interferes with the very parameters for the existing order. In the first case, being incapable is a deficient potentiality, a dunamis that stagnates, not culminating in the energeia of historical emancipation. In the second, it leaves ajar a window of freedom—unto death—irrespective of physical deprivations, short-term pragmatic calculations, and practical roadblocks.

More than a century after its release, *Reflections on Violence*, with its advocacy of a proletarian general strike, has been classified among the blind alleys of radical political philosophy, a lost cause if there ever was one. But, precisely, lost causes are the repositories of political energy insulated from the potentialities of the foreseeable and the achievable. Upon a careful reading, the book begins to sparkle with unmatched perspicuity pertinent to our own situation. For instance, Sorel presages the rhetoric of the European Union, drained of the last vestiges of political energy and going down the path of entropy. "Two accidents alone," he writes, "would be able to stop this movement: a great foreign war, which might reinvigorate lost energies . . . or a great extension of proletarian violence." Hence the rhetoric of popular orators: "European peace must be maintained at any price; a limit must be put on proletarian violence."[12] These are, indeed, the recipes for the European Union's suppression of inner tensions and crises, notably in the relation of its debtor and creditor member states. The EU's technocratic leadership dreams of an apolitical world, unconditionally divested of political energy, pacified rather than peaceful, free of violence but full of force. The threat of terrorism and the discontent of un- and underemployed youths are the new kinds of "accidents," heralding a rude awakening from this dream.

I take Sorel's brilliant theoretical separation of force from violence to be his most significant and abiding contribution to political thought. In a nutshell, "the object of force is to impose a certain social order in which the minority governs, while violence tends to the destruction of that order."[13] Force is the potency of potentiality, a form of dunamis that freezes in the work of politics at the behest of the status quo. Violence is the irruption of energy. Not necessarily tied to an active doing, it rejects the coordinates of an order imposed by force and sets the dream of self-organization to work. Odd as this may sound, violence cannot be forceful on pain of colluding with the operations of the very regime it casts off. It abnegates force not in order to pacifistically accept "the way things are" but to change them from the ground up. Its energy is the energy of the end, not to be confounded with a cessation, say, of individual or institutional existence. Violence comes to us from the future of freedom and fulfillment; force is emitted from the past of domination and illegitimacy. Revolutions end in a fiasco, in part, because sooner or later they diffuse violence in force, engaging the regime they fight against on the very terms it dictates.

The political energy supply thus lies in the future, not in the past of forgotten origins and dusty foundations. Moreover, it is only with regard to the future and its temporal exteriority—its extemporaneousness—that the past can be reenergized. As an alternative to the extraction of political energy from the kernel of the founders' intent, presumably buried underneath strata of history, revolutionary violence locates its motivations on the surface of "contemporary myths," such as that of a proletarian general strike, which "lead men to prepare themselves for a combat which will destroy the existing state of things."[14] A long way from utopian pledges that activate reformist attitudes, the promise of "powerless" violence, contesting the force of domination with the actuality of an end rather than with a counterforce, supplies political agents with energy above and beyond the potentialities inculcated in the current state of affairs. The general strike sets one concrete precedent for this maxim; another is Gandhi's so-called nonviolent resistance, which, in the best traditions of powerless political energy, is also eminently violent, though not at all forceful.

Like Aristotle's energeia, the violence of the proletarian general strikers and of Gandhi cannot be explained in the language of instrumentality, of means selected to achieve some intermediate ends. Benjamin consecrates this exceptionality in his essay "Critique of Violence" by characterizing Sorel's stance as that of "pure means," reine Mittel, conceptually cognate, we might add, with pure ends. "For it [the proletarian general strike] takes place," he explains, "not in readiness to resume work following external concession and this or that modification of working conditions, but in the determination to resume only a wholly transformed work, no longer enforced by the state, an upheaval that this kind of strike not so much causes as consummates [*nicht sowohl veranlaßt als vielmehr vollzieht*]."[15] Benjamin's choice of verb is hardly haphazard: the effect of the proletarian general strike is to consummate or fulfill (*vollziehen*) a complete transformation of work, its conditions and procedures, regimentation and execution. What takes place in an indefinite suspension of labor is a global work on work, which, well beyond the scope of an adjustment ("this or that modification of working conditions"), demands justice. The overall backdrop for this demand, however, is not work, ergon, but energeia, the place of all consummation and fulfillment.

With Benjamin, who intersperses the conclusion to his "Critique of Violence" with enigmatic references to divine violence, we have inched toward the messianic connotations of energetic powerless in political theory and praxis. The weakness of God in Christian theology was vulnerable to Nietzsche's challenge, according to which it was a reactive and surreptitious permutation of the will-to-power in the guise of the will's—and power's—self-denial. How does this challenge stand with the proposals launched by Sorel and Benjamin? Reactive inactivity would be applicable to limited strikes that, in quid pro quo "collective bargaining," seek concessions from the capitalist, still admired as powerful and active. In a general proletarian strike, on the contrary, powerlessness becomes active over and above the relational differentials of activity and passivity, when the strikers declare their absolute separation from the exploitative other and destroy the old bourgeois order as a consequence of their self-affirmation in the face of brutal physical and economic necessity, the "laws" of nature and of capital. The energy of the end they draw upon is apocalyptic. They receive it from a brush with mortality, in the same sense that Hegel's slave is liberated from his subjugation to the master by being shaken to the core with the fear of death or that Heidegger's Dasein changes its attitude toward life by lucidly countenancing its end in being-toward-death. Because it declines to participate in the politics of small tricks, to resort to a phrase Lenin favored, revolutionary powerlessness is neither reactive nor reactionary. So energetic and energized it becomes that it transmits its charge on to the adversary: "in an indirect manner it can operate on the bourgeoisie so as to reawaken them to a sense of their own class interests."[16]

***

Were I asked to summarize the take-home lesson of *Energy Dreams*, I would use a heuristic device of traffic signs. Two of them: *Dead End* and *One Way Street.*

Dead End. The end itself is dead, suffocated under a sky-high pile of means, *and* we have gone astray, getting caught at an impasse. We have been looking for energy through distorting prisms and in all the wrong places—in the depths of the earth, not on sun-soaked surfaces; in the everlasting idea and an eternal divinity, rather than finite thinking and a mortal god; in an endless accumulation of power in place of a vigorous powerlessness. As though we have never heard, or have heard but not listened to, a word of Aristotle, who patiently recites that power-*dunamis* is not *energeia*. Regardless of the quantities of power you accumulate, you will not obtain the ultimate nonobject of your desire, the stuff of all your dreams.

One Way Street. Despite all the tantalizing options we are dreaming of in the rut of the Dead End, we have only one momentous choice and one chance, namely to reverse direction, to shrink back from the bottomless pit of pure potentiality, and, at last, to piece energy together out of the actuality we have neglected and the dynamism we have overindulged.

For millennia, we have been reading Aristotle backward and upside-down. This manner of reading *is* metaphysics. Hence the authors, including the classics, are not metaphysical; their readers are. We thought that the actuality of energeia was stifling and oppressive; we harbored the ambition to overturn its relation to dunamis, and, giving blank possibility a carte blanche, tried to buy our freedom with it.

Our conception of the state—presently, of transnational political institutions—is complicit in the dead-end fascination with potentiality, which is conflated with energy and amassed for its own sake, endlessly. Even if he treats power and energy as interchangeable, Vladimir Bibikhin is exceptionally lucid on this point, noting that the chief political goal is the same here as the means for the achievement of any goal: "power, the increase of power—economic, political, military—the growth of energy. . . . Thus, the possession of energy, of power, seems to become the end in itself: mere might, which is especially notable in standoffs between states. The state strives towards might."[17] For the status quo to preserve itself (say, through the use of force, as in Sorel's reflections), to guard against entropy and revolutionary violence, it is not sufficient for it to remain static, unperturbed and unchanged. Wedded to dunamis, it must constantly augment itself, expand, grow, gather more and more strength. Or, better yet, it operates on the assumption, which it has in common with the economic system shoring it up, that it can only preserve itself by growing, by increasing the virtual potentialities of its power, already well in excess of what is needed to pulverize the entire planet.

On the one hand, we have states and other institutions that, in a mad dash after energy, get a hold of power, of potentialities, in lieu of actuality. A swelling potentiality spills over into the growing void that, more and more murderously nihilistic, threatens to devour the remnants of actual existence. Terrorism, Rasmus Ugilt explains, is a steady companion of this political ontology, because its "energy" is pure potentiality, a potential menace "against everyone and anyone."[18] On the other hand, we have *activists* who fight for social, economic, racial, sexual, or gender parity and justice, but whose political activity is so vague that it uncannily mimics formal power-dunamis. Their action frequently lacks an object. (Something of the kind comes to pass in the academia with the comparably objectless designation *theorist*.) Esteemed in "progressive" circles, activism veers toward a reaction and reactiveness, even against those who sympathize with whatever its cause might be. So, to clarify things, we ought to ask: Who or what is activated in political activism? What is its relation to energy? What is energizing and who is energized? What does it set to work and how? Lest we sift through these queries, activism will be restricted to a working without a work, hostage to means without an end, a sheer acting out, or a pale copy of capitalist accumulation.

How, then, is it possible to reinvigorate political energy? We have received some clues from anarchism and anarcho-syndicalism, leading us to surmise that powerlessness, impotentiality, and weakness are more akin to energetic politics than the accrual of power, potency, and force. It is difficult—and for good reasons!—to condense these hints into a political program, as Benjamin has observed with reference to Sorel's general strike. A program, after all, presupposes a linear realization of potentialities toward a given end, whereas energy feigns a standpoint at the end, from which to scrutinize and engage in existence, itself necessarily a coexisting. Therefore, "Sorel rejects every kind of program, of utopia—in a word, of lawmaking—for the revolutionary movement."[19]

Outlining an alternative program is pointless, but at least we can tweak political concepts so as to impart to them an energy at odds with the empty potentialities of power. An obvious place to turn to as we scan for guiding lights is Hegel's political philosophy. Condemned for his "totalitarianism" by twentieth-century liberal zealots, the German thinker fell victim to the same fate as Aristotle, and for much the same reasons: he was submitted to an unabashedly metaphysical reading. In the beginning of the section on the state in *Philosophy of Right*, Hegel pithily writes: "The state is the actuality of the ethical idea [*Der Staat ist die Wirklichkeit der sittlichen Idee*]."[20] The conventional interpretation is that Hegel reckons the nineteenth-century Prussian nation-state to be the culmination of Spirit's history, the highest form of political and metaphysical organization. Yet, as we have learned, Wirklichkeit, translated as "actuality," is Hegel's adaptation of the Aristotelian energeia. The state is, then, contrary to the mummification of the rational in the actual, the energy of the ethical idea, the idea's being-at-work, or setting-to-work, while also being at rest. The political unit is substance and subject, "the substantial will," *substantielle Wille*, or, in our terms, the working and the work of the ethical idea. As the social world's self-relation, replaying the logic of the unmoved mover, political energy thinks and knows itself, *sich denkt und weiß*, and, insofar as it does, fulfills (*vollführt*) itself.[21]

Energetic fulfillment is what Hegel intends when he talks of actuality and, chiefly, the actuality of "concrete freedom" ("The state is the actuality of concrete freedom [*Der Staat ist die Wirklichkeit der konkreten Freiheit*].")[22] Concrete freedom is the reconciliation of universal and particular interests, as well as of movement and rest that no longer cancel each other out. It announces release, an energetic discharge that, rather than deplete, lets blossom the energy it expresses.

In Hegel's nineteenth century there was no doubt that the nation-state was the most perfect political organization, which is why he dreamed of it as the concretion of rational actuality, the utmost energy of the ethical idea. Already with Marx, however, the universality of the state was superseded, while in the twenty-first century states are materially divested of their sovereignty, for better or for worse. We might say that the word *Staat* temporarily occupies the otherwise empty place of the signifier for the political community and that, if need be, it can be replaced with other, more historically appropriate designations of living together. Such changes would not drastically affect the sense of political energy that guarantees concrete freedom, the reconciliation of movement and rest, or the self-relation and self-accomplishment of the many. Indeed, the beauty of this energy is that, by laying the stress on the fulfillment of potentialities and the common, it applies both to anarchism and to state theory.

Before discarding the word and the concept (if not the institution) altogether, we ought to consult the semantics of the state, or Staat, relevant to political energy. Derived from the same root as *stasis*, the state is the movement of rest or the rest of movement. In conjunction with the more abstract status quo, stasis slips unrest into the formation of order—hence the debacle of revolutions that, aspiring to regime change, only further energize the state or that, taking pains to do away with the state form, mistake themselves for the external negations of order, with which they are embroiled in the overarching logic of stasis. That is why, also, Schmitt is wrong in his 1933 essay "*Staat, Bewegung, Volk*" ("State, Movement, People") to assign to the state the static reality of the political apparatus, while locating dynamism in the popular movement, serving as a mediation between this apparatus and the people.

Between the static and the dynamic, energy itself is mislaid. A good exit strategy from this crisis, for those who are looking for one, is not to smash oneself against the machinery of the state, which, particularly under the aegis of capitalism, feeds on revolt. What is required is a change in the energy of the state, steering the balance of movement and rest in the direction of Hegel's concrete freedom, the nonideological coincidence of the universal and particular interests.

If, in *Phenomenology of Spirit*, state power (*Staatsmacht*) decays into an "*abstract universal*,"[23] the state itself is not at fault. What is blameworthy is its combination with the power (*Macht*) that siphons political energy into potencies and potentialities. Hegel recommends moderating abstract universality with the concrete individuality of the monarch and resuscitating the political sphere through the hypostasis of sovereignty in the sovereign. The Miller translation turns this into an issue of energizing the political: when it still persists as an abstraction, "state power has yet to be energized [*begeistet*] into a self."[24] The verb *begeistern* literally means "to bespirit" or "to inspire," though its more common renditions are "to enthuse," "to beguile," "to enthrall." At any rate, the link between being energized and being inspired or in-spirited is revealing: in it flickers something of energeia retained by the Hegelian *Geist*.

Just as the state is jettisoned by everyone, from the advocates of transnational governance to the radical left, so the very concept of sovereignty is generally spurned in our age of dispersed networks and microlevel power-knowledges. The question that remains unasked among these knee-jerk reactions to a term tainted with theological and absolutist heritage is: What are the outlines of energetic, rather than dynamic, sovereignty? To venture into a largely uncharted territory, I would suggest that, in keeping with political energetics, the sovereign is the one—or more, or less, than one—in whom power rests. I am not, by any means, advocating replacing democratic rotation with dictatorial stagnation, the ceaseless movement of power's circulation with torpor. Energetic rest, as we know, is consistent with the fullness and fulfillment of activity, which, outwardly impassive, is more active still than the fumbling of means without ends, of dunamis unmoored from energeia. The sovereign in a condition of energetic rest is not *invested with* power. The political energy of sovereignty is such that nothing changes depending on whether or not sovereigns *have* the reins of power, which they do not lack and therefore do not crave. Stamped by supreme powerlessness, this sovereignty is an energy in excess of potency and potentiality. You will probably smile when I point to the most influential example of powerless sovereignty in the history of Western thought: Plato's philosopher-kings.

—— ∞∞∞ ——

In addition to dreaming about the fusion of theory and practice in a perfectly governed polity, Plato had a thing or two to say about the protocols of a proper political conduct. When the Athenian from his *Laws* pronounces himself on the subject of temperance, he stipulates that "judges while on duty [δικαστὰς ἐνεργοὺς ὄντας] should not taste any wine at all" (674b). The word rendered in English as "on duty"—or, sometimes, "in effect"—is *energous*. Remarkably, the sole mention of energy in Plato's texts, before the Aristotelian codification of energeia as a philosophical concept, is political. The officials in their offices, magistrates at work, should be altogether sober—that is the requirement for the optimal discharge of their functions. Their energetic performance demands utmost lucidity, discernment, sound and unclouded judgment. Being "swayed by right reason and the law [τοῖς νοῦν τε καὶ νόμον ἔχουσιν ὀρθὸν]" is inconsistent with inebriation (674b), and neither is the energy, the political activity and actuality of representing an institution, temporarily invested in public officials.

Technocracy, which presides over today's political landscape, severs the old ligature that united politics and energy. In Western countries, it no longer matters who is actually in office. With minor variations, the activity of officials is in line with the exigencies of global financial institutions and the flows of capital. National and transnational political phenomena are presented as the workings of a neutral and gigantic machine, with an implacable program encrypted by the market and a necessity that overrides public discussions and parliamentary debates. Technocracy's energy dream is to revamp the role of politicians, who would be not so much machine operators as the conduits for its operations, the oil in its gears. Under these circumstances, the inebriation of public officials (memorably, the late Russian president Boris Yeltsin) would probably contribute to their politicization rather than stand in the way of lucid judgment and correct reasoning, praised by Plato. Their drunken conduct would foreground the hollowness of diplomatic protocols that obfuscate a shameless promotion of economic interests by and for the wealthy elites.

Technocratic neutrality falls flat when it comes to the fantasy of pacification, the absolute withdrawal of political energy. Outside the sterilized space of institutional politics, negative energy swirls without a chance to be objectified in an actual structure, shape, or process. In psychoanalytic terms, this energy is unbound, uncathected, and therefore profoundly melancholic, an "open wound" that further impoverishes the public sphere. Political melancholy is not only symptomatic of postrevolutionary conditions, as Artemy Magun has astutely argued,[25] but also of any situation where the existing institutions are incapable of binding group affect. As a result, freely mobile energy spills into a general malaise, interspersed with vague complaints about the status quo and the widespread corruption of politicians. Voting becomes largely reactive; "acting-out" in its political decisions, the electorate swings from center-left to center-right parties (or vice versa) and casts ballots *against* whoever is in power. Nihilism is a passive-aggressive countenance of this energetic lopsidedness, of acting without actuality both in politics and outside it. Terrorism is its active-aggressive manifestation.

—— ∞∞ ——

Approaching the topic of political energy, the first impulse is to smuggle the concepts, correlations, and axioms from the physical to the social sciences. We might ask, for instance, how the laws of thermodynamics bear upon the fluctuations of force, causing segregation between and aggregation within antagonistic human group formations. Needless to say, this methodology would succumb to crude scientific determinism.

Curiously, French sociologist Gustave Le Bon developed his researches in the inverse direction. After the release of his popular treatise on "crowd psychology," *The Crowd: A Study of the Popular Mind* in 1895, he became fascinated with physics and published *The Evolution of Matter* and *The Evolution of Forces* in the first decade of the twentieth century. According to Le Bon, the crowd is the matter from which social and political entities can be sculpted, and matter as such is a temporary conglomerate that undergoes gradual dissolution and decay. The crowd's transitory "collective mind" explodes with violent force and is roused to action as the conscious personality of its participants momentarily vanishes.[26] With the same ease, it falls apart again into its individual constituents. Contravening the first law of thermodynamics, matter and energy do not enjoy "the privilege of immortality": they "also must enter into the cycle of things condemned to grow old and die."[27] Anything but a peaceful dusk of entropy, the splitting of matter at the sub- or intra-atomic level results in the liberation of large quanta of energy stored in the atom, its "colossal reservoir."[28] Implicitly parallel to the breakdown of individual consciousness and the explosion of collective unconscious action, Le Bon's physics is an extension of his social theory, not vice versa.

Several decades later, Carl Schmitt will infer political energy from the analogous physical phenomena of pre-Einsteinian physics where matter and energy are treated as separate. Although he does not overtly use these terms, the political for him is not matter but energy, not substance "but only the degree of intensity [*Intensitätsgrad*] in an association or dissociation of human beings." It follows that, as a pure intensity or a differential of forces, the "political can derive its energy from the most varied human endeavors, from the religious, economic, moral, and other antitheses."[29]

Schmitt's version of political energy is a dynamic form, the dynamism of formation, a potentiality never satisfied with any concrete instantiation and independent from its material substratum. So long as there are two planes that, in friction with one another, clash with ample force, political sparks will fly. Not attached to a single substantial domain, the political thrives on the energy of opposing surfaces that Schmitt, uninterested in their actual content, defines as *friends* and *enemies*. It is utterly superficial, free of a deep essence, animated from the outside by the exteriority of enmity (even if the enemy is found within a given political unit) that puts one human association on a collision course with another. Political energy is electric, with the positive and negative charges of friendship and enmity pulling on and pushing against each other in a previously neutral, apolitical field they perturb.

But the physics of political energy is, by far, not the whole story. To the extent that it maintains sovereignty, every political entity retains the right to "decide on the friend-and-enemy distinction and, if necessary, make war."[30] Concurring with the assessment that the distinction itself is squarely within the purview of the (modern) physical outlook on energy does not mean that the decision on this distinction pertains to the same outlook. Much has been made of this sovereign *decisionism*, which, with its unruliness, smacked of the theological, first and foremost in Schmitt's own estimation. But the assumption *whatever does not obey mechanical, electromagnetic, or atomic laws transposed onto political life is irrational* is severely flawed. Science or magic: that is a false alternative. Sovereignty is experienced as a miraculous irruption, *sensu stricto*, from within the scientifically determinist paradigm of human coexistence. Our bankruptcy is glaring the moment we prepare to pay for sovereignty with the coin of natural or social-scientific law for "heads" and supernatural power for "tails."

With few exceptions, we are comfortable with the attribution of events to physical causality or, when its laws are inapposite, to metaphysical or theological causes. But, in this default setting of our thought, we also neglect another energy, neither entirely immanent to nor transcendent vis-à-vis a given order. The smooth grid of legality functions in a way akin to the cause-effect relations in classical physics. However, the energy of legitimacy is irreducible to the causal operations it sets in motion in those societies that, at least nominally, respect the rule of law. Sovereign energy, yielding decisions on who is a friend and who is an enemy, or, *priore loco*, deciding to decide on the exception, puts a particular political order to work, activates or actuates it, while not being explicable on the terms of that which it has activated. The definition of the sovereign as "the one who decides on the exception [*wer über den Ausnahmezustand entscheidet*]" essentially conveys that an exception decides on the exception,[31] conducting a doubly singular energy, which cannot be squared with the ideology of formal equality prevalent in liberal-democratic societies. Perhaps only quantum physics can be still in sync with the zero point of sovereign energy, soliciting an infinite approach, such that the singularity of the sovereign exception would function as a black hole of political thought.

His vehement Catholicism notwithstanding, Schmitt finds himself in agreement with Kierkegaard, a "Protestant theologian who demonstrated the vital necessity possible in theological reflection,"[32] on the method of thinking the general starting from the exception. He repeats, with pleasure and approval, the words of *Repetition*: "the general is not thought about with passion but with a comfortable superficiality. The exception, on the other hand, thinks the general with energetic passion [*Die Ausnahme dagegen denkt das Allgemeine mit energischer Leidenschaft*]."[33] Passion and pathos, passivity and energy, can be reconciled: suffering through and lingering with singularity yields more energy than an active generalization ever could. Sovereign decisions and thinking about sovereignty are illegitimate, unless they *experience* the exception, are laden with its unbearable weight, and receive the energy oozing from its outer border or limit—the *peras* of *ex-perience*. The sovereign is not an active, original, ever-present source of politics under the unwavering influence of metaphysics, a power supply or a transcendental cause that apportions meaning, life, and death. Sovereign energy, rather than power, revolves around passivity beyond the binary of activity and passivity, a being-excluded that will, later on, resonate with the exception, upon which decisions will be passed. The sovereign is decided into existence by the decision on the exception, the energy of legitimacy swirling in a circle foreign to the thinking of legality.

Schmitt's constitutional theory is also comprised of two energy regimes, two political tectonic plates, roughly matching the physical and biological energetic models. The first relies on a comparison between constitutionalism and entropy, the original decision on the form of political existence forfeiting its energy in its very outcomes. Formalization is the precondition to a minimally organized existence or coexistence, as well as a precursor of an entropic finale, "the dissolution of the unified constitution into a multitude of individual, formally equivalent constitutional laws."[34] What for Hegel was a necessary moment of mediation, of a simple and abstract unity losing itself in the many only to find itself in a more concrete, determinately negated, shape, is for Schmitt a lamentable relativizing of political life. The aporia of constitutional energy is that it immediately wanes and becomes impossible in everything it makes possible, evaporating from what it sets to work. The force of law—diffracted into many legal statutes, fundamental as they might be—robs political energy of its vitality. The dunamis of legality offsets constitutional energeia.

The other energy regime is evident in the "absolute concept of the constitution," which combines "the concrete manner of existence that is a given with every political unity" and "the principle of the *dynamic* emergence of political unity, of a process of constantly renewed *formation* and *emergence* of this *unity* from a fundamental or ultimately effective *power* and *energy*. . . . The constitution is the active principle of a dynamic process of effective energies [*das aktive Prinzip eines dynamischen Prozesses wirksamer Energien*]."[35] In contrast to the fragmented relative constitution, the absolute is the constituted, the constituting, and the relation between them: the work, the workings, and the enworkment of political life. It mobilizes rest and movement, being and becoming, the existence of a polity and its formation, unity and unification. "The sharp separation of the static and dynamic has something artificial and violent about it," Schmitt adds.[36] Or, in our terms, the passage of activity into actuality, of working into a work, is integral to energy. Every constitution that *is*, that *brings into being*, and that *becomes* recreates itself in a struggle against entropy, similar to a living being, whose aging, nevertheless, is not at all equivalent to entropic loss.

The homology of political units and organisms latent in the "absolute concept" of the constitution is, of course, not new. It harks back through Hegel and Hobbes to Aristotle and Plato, who kept abreast of the *life* proper to political entities. Schmitt's innovation lies in juxtaposing the physical-inorganic and biological-organic types of energy in a constitutional framework that involves both and thus obliges us to perform an arduous balancing act, juggling a multiplicity of formalized laws and political existence, static statutes and dynamic decisions, alliances and warring factions. Lest we get carried away, the separation between these facets is "artificial and violent," Schmitt reminds us. Each extreme is indispensible to a political energeia that, in acting, counteracts, divides against itself. The enworkment of politics unleashes intense antagonisms, distilled from the antitheses that structure diverse spheres of life. But what allows politics itself to be political if not its revolt against itself, its becoming an enemy unto itself? Depoliticization is the highest possibility of the political, the apparent diminution of its energy—its greatest surge. To speculate: only Schmitt's fierce aversion to liberalism could have blinded him to this blatant dialectical ramification of his political thought.

The energy-saturated subtext of Schmitt's writings opens a promising avenue toward understanding the issue of political representation. In keeping with the entropy hypothesis, representation lives on usurped energy, constantly tapping into the hidden source of legitimacy, of who or what is represented in it, be it "the people" or the monarch's absolute will. The further away the representing from the represented, the less energetic charge does representation carry, and the more it is subject to the forces of entropy, cooling down, and decay. Qualifying democracy with the adjective *representative* is more harmful and apolitical of a gesture than bestowing upon it the title *liberal* or even *parliamentary*. Such a designation means that this regime subsists on usurped energy, lacking a political charge of its own. If the impetus of democratic institutions had initially been a vigorous discussion, wherein friend-enemy relations had been transformed into a nexus of debating adversaries, that (already suspect) stimulus animating them has now faded, giving way to a lackluster, tired, contentless representation, predicated on a series of false identities between the representatives and those they represent.[37]

"The situation of parliamentarism is critical today," writes Schmitt in *The Crisis of Parliamentary Democracy*, "because the development of modern mass democracy has made argumentative public discussion an empty formality. Many norms of contemporary parliamentary law, above all provisions concerning the independence of representatives and the openness of sessions, function as a result like a superfluous decoration, useless and even embarrassing, as though someone had painted the radiator of a modern central heating system with red flames in order to give the appearance of a blazing fire."[38] In other words, the dynamic formation of political energy has been substituted with the static form ("an empty formality") of apolitical institutionalism. The last vestiges of combat, if only on a verbal arena, are eliminated, and whatever remains of them is purely ornamental, much like the visual representation of fire Schmitt invokes. Representative parliamentary democracy is in crisis, precisely because it has drifted too far from the source of the political, from the boiling energy that drives friend and enemy groupings. The crisis, then, is not one of democracy "itself," nor of representation "itself," but of the energy that has been cut from that which it energizes, resulting in a situation where the latter moves by inertia and obeys nothing other than the law of entropy.

On the one hand, Schmitt seems to believe in the metaphysical fiction of energy as a purely living source, the transcendental signified of who or what is represented animating the rest of the political field. The same applies to his implicit theory of knowledge, which, in a signature gesture of Platonism, prioritizes the fire itself over its drawn representation (though the *idea* of fire might be open to polemical—that is, political—contestation). Hence the critique: representative democracy is a democracy of borrowed energy, this borrowing—the representation—made primary and, indeed, definitive of a regime that bears its self-alienation as a scepter of its power.

But, on the other hand, the political as such is an essentially impure (I would even say, "essentially nonessential," i.e., existential) source, its origins indefinitely displaced and buried in nonpolitical domains that may, all of a sudden, become politicized. Always on the verge of depoliticization, political energy reverts to a specific domain whose usual structure it has temporarily perverted; in fact, it *is* but a perversion without a set standard of normalcy. Its inseparability from the finite experience (the undergoing, the *pathos*) of the political means that it is in equal measure receptive and active, including with regard to the representations presumably based upon it. The expression of its vitality in an oppositional relation, let alone in the objective or substantive "works" it undergirds, and its entropy are of one piece. Representation is this simultaneity of the expression and entropy of that energy which is not formalized from the outside, the energy that gives itself both substance and form as an effect of its desubstantializing intensities.

The strategies Schmitt favors with the view to saving political energy (especially—though he doesn't see it this way—from itself) are multifaceted. And most of them are discernible in his book on Roman Catholicism and the early essay "The Visibility of the Church." For one, energy may be kept whole to the extent that it is neither expressed nor represented nor maintained in a secret reserve: "To every great politics belongs the 'arcanum.'"[39] Strangely enough, the political energy that eludes expression and representation is not simply sidelined, no longer putting anything to work. On the contrary, it animates politics all the more vigorously while staying withdrawn and therefore unrepresented, albeit not unrepresent*able*. Harking back to "the most extreme invisibility" of God (57), political arcana are not only energy detained in potentia but also the energizing factors, the conditions of possibility for politics removed from the field of political visibility.

A related strategy is to afford energy a certain degree of self-expression and self-representation, provided that it does not swerve too much from the expressing and representing core. That, in my view, is the basis for Schmitt's famous concept of "representation from above." Just as the pope is the vicar of Christ on earth, and just as the visible church is a sign for the invisible spiritual realm ("*in* but not *of* this world"; 52), so a political representative is the temporary concretion, the congealment of political energy that lends her or him legitimacy and "special dignity" (21). The swerve of this energy from itself is minimized thanks to the restriction of representation to expression, which Schmitt links to personification both in *Roman Catholicism* and in his study of Thomas Hobbes: "To represent in an eminent sense can only be done by a person, that is, not simply a 'deputy' but an authoritative person or an idea which, if represented, also becomes personified" (21). While clearly inherited from the doctrine of divine incarnation, the personalism likewise prevalent in Schmitt's theory of sovereignty is a particular case of representational expression. Here the integrity of political energy is ensured by articulating its verbal and substantive modalities, its working and its work (ergon), insofar as the former generates the latter in its actuality. The representative, in turn, becomes a mediator in this process internal to energy: a medium, through whom, neither immanent nor absolutely transcendent to the political order, energy conducts commerce with itself.

Rather than "personification," *Personalisierung*, what we are dealing with when it comes to the representational self-expression of political energy is *figuration*, rooted more in the world of Baroque allegory than in that of Patristic theology. According to Schmitt, "the Catholic Church is the sole surviving contemporary example of the medieval capacity to create representative figures—the pope, the emperor, the monk, the knight, the merchant. ... It stands so alone that whoever sees therein only external form mockingly must say it represents nothing more than the idea of representation" (19). Of course, the "external form" of figural representation is the expression of the inner form, brought forth by the energetic content itself. In the dialectics of political self-expression, developed by Schmitt with little reference to Hegel, figuration is only a temporarily detained *con*figuration of energy that attends its own representation. The dissolution of figural thinking in modernity, its decline in the face of rational abstraction, portends the dissipation of energy still capable of cohering into recognizable clusters of which everyday existence is composed. As a rule, with the eclipse of figuration in politics, political existentialism becomes untenable or, when still espoused, turns into a reactionary and markedly irrational, atavistic caricature of everything that is anathema to disembodied abstract rationality. Political reality unfigured is political reality disfigured and resistant to representation.

Whatever the strategy Schmitt adopts, he is reluctant simply to embrace the other extreme relative to economism, which stresses total visibility, materiality, and the techniques of administration. He is, emphatically, not an advocate for sheer invisibility, a purely spiritual refuge, or otherworldly mysticism. Whether in the occluded form of the arcanum that shapes the political field or in the self-expression of figural representation, the visible and the invisible are reconnected in an attempt to establish a precarious energetic balance conducive to (public) life. Political energy, alongside other kinds, is reducible neither to shapeless flows that, forever potential, lack even momentary actualization, nor to institutional structures, within which these flows are deeply frozen. It is, rather, an ongoing interplay between these two moments, between the invisible workings that belong together with faith and the visible works (58), between the animating and the animated that, as phenomenologists correctly note, needs to be incessantly reanimated, put back into play, or set to work. Stated in a somewhat unorthodox manner, that is the main lesson of Schmitt's writings on political theology in general and on Catholicism in particular.

As an astute reader will realize, the critique of representation exposes Schmitt to a barrage of deconstructive accusations, including, in *The Politics of Friendship*, by Derrida himself. On the whole, the German jurist stands accused of a not-so-secret allegiance to the metaphysics of presence, responsible for his predilection for some types of representation over others. This "easy" interpretation is faulty not only because it overlooks the originary impurity of the political—its bastardly derivation from every conceivable domain of human activity—but also because it misjudges Schmitt's self-positioning vis-à-vis the extremes of materialism and idealism, the relentless presence and visibility of things and the invisibility of "spiritual realities." The status of the Church and, therefore, of the political form Schmitt aspires toward is that of a trace, the staple figure of deconstruction denoting a present absence or an absent presence, the trace of another world or another time in this one. Any representation worthy of the name holds onto the figureless figure of the trace, politically and ontologically.

To choose presence over absence is to embrace economism; to opt for absence over presence is to fall into the clutches of private religion (read: Protestantism), which is actually not so different from economism, merely distorted and disavowed ("Historically considered, 'privatization' has its origin in religion. . . . Private property is revered precisely because it is a private matter"; 28). None of these alternatives appeals to Schmitt, and none is compatible with representation: the first, because there is nothing to represent where everything is ideally present; the second, because excessive "privatization" does away with language as such. The bridges and mediations he tries to build between these two abysses are not monumental, substantial structures. They are traces and tracings, fragile figures that crop up and disband in a historical instant.

Important as the treatment of representation is in and of itself, it is especially valuable for what it says about political energy as *energy*. As we have seen, this concept, too, has often been accused of belonging to the pantheon of notions that—more or less interchangeable with one another—buttress the metaphysics of presence. But Schmittian political energy (or, better, the political qua energy) is itself a trace of other domains of human activity it polarizes, of substance as well as subject, of qualitative distinctions and rigid systems of classification. Without identity, conflating even the difference between the static and the dynamic, *energy is a trace*. Beyond the immediately political applications of this conclusion, we might begin to sense its tremendous reverberations in ontology *after* the end of metaphysics.

6

PHYSICAL FANCIES

The current businesslike weltanschauung of science, with its rigid specialization and division of labor, successfully convinces us that scientists do not dream. More and more, they function as auxiliary instruments in the acquisition of knowledge or, at best, its managers. "Fundamental" physicists have probably dreamed more, and more daringly, than the rest of their colleagues, because, like philosophers, they have questioned the certainties of time and space, matter and mass, imagining alternative realities inaccessible in the world handed over to our direct experience. That they are less valuable on the academic marketplace than "applied" physicists is a sign of the times.

Scientists are discouraged from dreaming, least of all about that which deviates from the stipulations of economic efficiency and profitability. The wings of their imagination are clipped, on the one hand, by the techno-bureaucratic (or, indeed, technocratic) procedures that predetermine the outcomes of their research and, on the other, by their own disillusionment with the disastrous consequences wrought by "applied sciences." The effects of twentieth-century atomic physics on the world—from Hiroshima and Nagasaki to the Three Mile Island and Chernobyl accidents, as well as, now, Fukushima—are especially disheartening. As Valentin Borisevich, former laboratory head at the Institute for Nuclear

Energy of the Belorussian Academy of Science, confided in Nobel Prize winner Svetlana Alexievich:

> Three *kopeek* [cents], according to our school teacher, contained enough energy to put to work an entire power station. . . . The cult of physics! The time of physics! Even when Chernobyl exploded. . . . How slowly were we parting with that cult. . . . The scientists who arrived at the site of the reactor did not even bring their shaving implements, as they thought that they were coming over for a few hours. Only for a few hours, even though they were told there was an explosion at the atomic power plant! But they had faith in their physics; they belonged to the generation of that faith. The time of physics came to an end in Chernobyl.[1]

The energy dreams of physicists in the last century were catastrophic. Practically speaking, they were the deadliest in human history thus far. The self-styled peaceful applications of atomic physics spawned more victims than its military utilization. For the first time, the technological means for an instantaneous annihilation of the entire planet were acquired, even if the idea itself had predated its plausible realization. Contemporary physics is finally in a position to bring the ancient dark vision of metaphysics to fruition: to incinerate the earth in the split of a second, ridding all existence of its materiality, its heaviness, and converting its mass into mobile energy in a scientific *imitatio* of pure spirit. To put energy to work in a manner so total and ineludible that it would extract the hidden kernel from everything and everyone by means of the division of (what, until relatively recently, has been considered) the indivisible. To actualize the absolute deactualization of *what is*.

We ought to work through, in the psychoanalytic sense of this expression, such lethal energy dreams, a project that will take untold generations, seeing how ingrained the metaphysical-physical fantasy is in our collective psyche. But abnegating *any and all* dreams, including those that yearn for a nonviolent energy, cannot be justified by the past failures of imagination to respect the actuality of existence. Nihilistic resignation, a dearth of energy in the thinking or dreaming of energy, testifies

to our acquiescence to the old, harmful, and unsustainable ways of procuring this evanescent object-subject of desire. Not dreaming is falling
into the most terrible nightmare yet.

Recent discoveries in theoretical and applied physics can bring about
extensive destruction, bewilderingly inspired by the ancient dream of
energy's indestructibility. Einstein's famous formula amends the law of
the conservation of energy to implicate the conservation of mass, which
is equally uncreated and cannot pass away. Relativity theory, after all,
interrelates matter (mass) and form (energy) in a hylomorphic assem-
blage of $E = mc^2$. That said, mass is not the same thing as matter. Although
there is generally no matter without mass, the latter is only a measurable,
quantifiable expression of the former. What Einstein articulates is form
and formalized matter, itself on the path to becoming pure form. (Relativistic mass has nothing to do with the mass of a body at rest, and
photons, or the particles of light, are altogether massless.) The matter
of physics is antimatter, entirely divorced from hylomorphic existence,
which can be stripped away, peeled off as if it were a mask concealing the
energetic core of the real. Energy-mass equivalence itself—that is to say,
as such—is highly destructive, to the extent that it literally mobilizes the
mass at rest, making it into an unimaginably mobile mass, into energy.
While energy and mass indefinitely remain, matter recedes, sucked as it
is into the black hole of the world, reconstructed from the ground up in
post-Newtonian physics.

Let us venture another formula: the degree of our captivation by the
law of the conservation of energy is inversely proportional to our commitment to the preservation of the lifeworld. As we reassure ourselves
that energy is anything but lost across all its transmutations, so we grow
indifferent to its momentary detention in any given hylomorphic structure. Naturally, then, energy is never actualized; it merely goes through
its kinetic, chemical, thermal, electric, electromagnetic, gravitational,
elastic, nuclear, and other instantiations without a shadow of fulfillment.
Construed by physicists, it is dunamis cut off from energeia, about whose
existence it suspects nothing. And the law of the conservation of energy
spurns finality, finitude—indeed, life itself—and so aggravates the crisis
of energy, which we have spotted in the opening pages of this book.

In late modernity, energy is virtualized. Thoroughly potential, it is now the reverse of Aristotle's initial conception. For the Greek master, energeia was not-dunamis; for us it has come to signify the not-not-dunamis. With its Midas touch, *our* energy forces everything solid to melt into thin air more effectively than capital, its economic camouflage. We do not need to study Hegel to understand that its unfulfillable abstract potentiality is devastating vis-à-vis everything that is. Yet, technically, "potential energy" conforms to a temporary actuality. Coined by nineteenth-century Scottish physicist William Rankine, this notion in thermodynamics evinces the energy *stored* in any physical system.[2] Stored energy retains the potential for conversion into other types of energy, participating in an endless chain of events that will never be actualized. According to the topsy-turvy world of physics, the energy contained in "actual" systems is "potential" because it detains the open-ended conversion process. In point of fact, energy is actual, if never actualized, strictly in a transition from one type to another, in an infinite exteriorization of what is stored within the limits of a definite shape. Philosophically adduced, it is only when it is not. And that, also, is the fate of the reality energy underwrites. Our existence and our world *are* only when they *are not*: on the threshold of nonbeing, denied the chance of energetic fulfillment, peace, and rest.

That said, we are reassured that nothing is ever lost, because all energy is conserved. The law of the conservation of energy is one of the most basic in physics. In the 1920s Niels Bohr attempted to challenge it by arguing that it "did not apply at the level of individual atomic events." He had to retract his hypothesis soon thereafter.[3] Codified by the German chemist Karl Friedrich Mohr in 1837, this law presents energy as an atemporal substance with epiphenomenal variations of heat, electricity, motion, and so forth. The questions it ignores are: In what, or across what, is energy conserved? Where is time itself accumulated? Is time yet another "container," shorn of beginning and end, wherein a fixed amount of energy is stored? Is it, despite this apparent openness, an overarching closed system, gathering in itself all the other closed systems of physics? And what, finally, is the difference between time, infinity, and eternity?[4]

Problem is that one of the most influential laws of physics elides the cobelonging, or—in Karen Barad's felicitous word—the *entanglement*, of temporality with that which is to be conserved. Just as there is no energy without mass (except for the borderline and debatable instance of photons), so there is no time without energy and no energy without time. We are prone to imagining temporality as empty, an arrow shot to infinity and, like energy and the world, knowing no fulfillment. But the ancient intuitions on the subject of the fullness of time are more accurate: they intimate (to put this anachronistically, in ultramodern terms) that, instead of arising in an empty temporal tunnel, events give birth to time through their singular energy-mass. One such event, typically thought of as The Event, was the Big Bang; another is the birth or formation of any being, living or not; yet another is the combination and separation of molecules or the quivering of submolecular, subatomic elements, never identical even to themselves. As Lee Smolin concludes, the reality of time, i.e., its effectiveness *simpliciter*, motivates changes in the physical laws of the universe.[5] (The first law to be submitted to the test of "real" time is that of the conservation of energy, which, precisely, leaves time out of the picture.) I take Smolin's comment as a call to arms, a plea for time, as much as for energy, to be liberated from the hold of understanding energy in terms of an immaterial substance, unaffected by anything that might betide its hylomorphic epiphenomena. "The reality of time" is time's actuation—its enworkment—both as substance and as subject. By resisting the virtualization of energy characteristic of late modernity, temporal "realism" cares for finitude and reconciles physics with life itself.

Whether we pose the question of matter using the tools of contemporary physics or the methods of ancient philosophy, this pseudoconcept turns out to be an obscure pivotal point of thinking about energy. Matter, in effect, has been persistently tinged, to the point of being saturated and defined by it, with a fair degree of obscurity, analogous to that shrouding the nonconcept of energy. Relegated to the sphere of potentiality by Rankine and reduced to mass by Einstein, matter connotes stored energy awaiting release. In the Aristotelian universe, it stands for energetic rest and actuality. Contradictory as these two enunciations sound, they unexpectedly share the presupposition that matter is the energy of energy needed for any putting-to-work; it is what puts to work the putting-to-work *in quantum huiusmodi.*

At another level, the incongruity between Aristotle and quantum theory is also more negligible than it appears. Energy rests in matter.[6] The opposite of lethargic repose, this is the height of activity, because matter itself is never completely at rest; it does not rest in itself, since it does not boast a "self," with a sealed and stable identity. Quantum matter and, by implication, quantum energy oscillate even in the case of a single electron, which is never purely single, never isolated, within the "fundamental discontinuity of quantum physics" that "disrupts the nature of difference." "The relationship between continuity and discontinuity," Barad continues, "is not one of radical exteriority but rather of agential separability, each being threaded through with the other. 'Otherness' is an entangled relation of difference."[7] What we perceive as an identity, down to matter or energy "itself," is the minimal movement of active, "agential separability"—in the first place, from itself. The energy of rest is the striving of matter to itself, its *nisus* or *conatus* that leaves sufficient room for physical fullness or fulfillment independent of form. It is matter's longing for itself, when nothing is missing, in the plenitude of materiality. That is the crux of Barad's "agential realism": *"All bodies, not merely 'human' bodies, come to matter through the world's iterative intra-activity—its performativity.* This is true not only of the surface or contours of the body but also of the body in the fullness of its physicality, including the very 'atoms' of its being."[8] "The world's iterative intra-activity"—is that not a contemporary designation for Aristotle's unmoved mover as the world's own self-relation?

Better than Nietzsche's will-to-power, the nisus of matter longing for itself encapsulates the being of becoming, which Heidegger painstakingly tracked in the writings of his predecessor. It traverses continuity and discontinuity, stability and change, by reverting back to the ancient energy of rest. Hegel knew that the principle of identity was the first to be sacrificed by anyone who wished to accompany motion and rest in thought. Not only his *Philosophy of Nature* but also his entire system commences there, with a rebellion against the abstract opposition of the two, against a way of thinking where "motion and rest are enunciated in accordance with the principle of identity: motion *is* motion, and rest *is* rest; the two determinations are external to each other [*Dies heißt nichts anderes als Bewegung und Ruhe nach dem Satze der Identität ausgesprochen: Bewegung ist Bewegung, und Ruhe ist Ruhe; beide Bestimmungen sind gegeneinander ein Äußerliches*]."[9] The main thrust of dialectics is to overcome this externality, to negate—and, in negating, mediate—the independence of each, achieving their interdependence. Energy is the summation of all the negations and mediations of motion and rest that give rise to the shapes of actuality, Wirklichkeit.

Matter's longing for itself, its nisus, is, for Hegel, unquenchable on the grounds of matter alone, which is essentially eccentric. "Gravity," he writes, "constitutes the substantiality of matter; this itself is the nisus, the striving to reach the *centre*; but—and this is the other essential determination of matter—this centre falls *outside it* [*diese selbst ist das Streben nach dem—aber (dies ist die andere wesentliche Bestimmung) außer ihr fallenden—Mittelpunkt*]. It can be said that matter is attracted by the centre, i.e., existence as a self-external continuum is negated. . . . The centre, however, is not to be taken as material; for the characteristic of the material object is, precisely, to posit its centre *outside itself.*"[10] In pure matter, the energetic nisus is deficient, in that it cannot be accomplished on the grounds from which it stems. Now the center always and necessarily falls outside matter because matter has no "self" to speak of; it is not for-itself, self-negated and mediated with itself. Its nisus is doomed to be an abstract potentiality, a dunamis that can shift into energeia (actuality, Wirklichkeit) solely outside matter, in the center that is both eccentric and ideal in relation to it. A place of fullness and fulfillment, the center where the material impulse can attain energetic rest, is categorically not material. In Hegel's thought, matter in general is not negatively mediated enough to have a self, an identity from which it could be liberated. It therefore, lacks an other, who or that (let us not rush to call this other *form*) would come about by way of negating that self. All this, together with dialectics tout court, presupposes that matter's eccentric simplicity needs to be overcome with the help of negative energy, destructive of something in matter in the name of its fulfillment. Hegelian energy is the liberation *from, of,* and *for* matter, which can neither be nor do anything for itself prior to achieving a self-mediated, self-negated ideality, shockingly less abstract than sheer materiality.

The dialectical energy of rest is viable on the condition that an identity is achieved, negated in favor of a self-identity, and ultimately placated in self-realized Spirit, wholly in and for itself, intending-acting upon itself in the circular closure of its nisus. *Voilà*—a physics painted in the image and at the behest of metaphysics, disavowing matter's own activity, activation, or putting of itself to work. It is, by the same token, a physics made inferior to biology: the energies of the living are, after all, truer to the dialectical ideal because more self-negated and self-mediated than those that traverse the field of "mechanics." The very scheme of *Philosophy of Nature*, proceeding from geometry via physics and astronomy to biology and meant to reflect the march of Spirit through a physicality that was initially other to it, substantiates this valuation. The disjunction between physical and biological energies is a haunting theme, and we will still revert to it. But rather than bid farewell to Hegel here, I want to reiterate a point from the beginning of this book, namely that his dialectics may be read as an extended meditation on the *en-* of *en-ergeia*, on the interiority of work or of the *at-work* proper to Spirit and making it what it is. It is this orientation toward energetic inwardness that impels the German thinker's desire to move past all kinds of exteriorities relative to the one and only interiority of *Geist*, the exteriorities where a work cannot be genuinely in or at work.

As a rule, metaphysics asserts its dominance over physics whenever material energy, movement, or striving is interpreted as a token for another reality, a higher and ideal Energy. In Hegel's system, the striving (nisus, *Streben*) of matter, be it organic or inorganic, portends the movement of Spirit on the basis of absolute rest. In Leibniz's thought, the physical nisus is the "force of nature implanted everywhere by the Creator [*vim naturae ubique ab Autore inditam*]."[11] Although, for all intents and purposes, "the order of things is not a matter purely mechanical, or purely metaphysical, but an intertwining of two 'kingdoms': the one . . . of efficient causes; and the other . . . of final causes,"[12] the two kingdoms and causes are not located on the same level. The entanglement of matter and spirit, each representing certain tendencies, forces, or energies, follows a hierarchical ordering of the signifier and the (transcendental) signified. The implication of the world's division into two "kingdoms" is that no meaning could ever germinate from material energy, save through the interference of Spirit. Consequently, energy itself has to be partitioned into a living force, *vis viva*, instilled by God in creation and a dead force, *vis mortua*, vaguely anticipating today's potential energy but incapable of crossing the threshold of actuality.

Having offered an olive branch to the ancients in a bid to rehabilitate final causes within the nascent worldview of modern physics that registers nothing but efficient causality, Leibniz did not resurrect Aristotle's word *energeia*. *Energy* will not make its discursive comeback until the middle of the nineteenth century, specifically, in the "energetics" of William Thompson and Rankine.[13] Before that, physics only knew the concept of force with its connotations of violence and potentiality. Leibniz, too, continued to rely on the Scholastic notions of force (vis) or striving (nisus), and he christened one of his projects "A Specimen of Dynamics," *Specimen Dynamicum*. This heavy legacy of Scholasticism explains the bifurcation of force into the living and the dead, replete with perpetual unrest, on the one hand, and the search for energetic fullness in God, on the other. Erroneously, Leibniz alleges that "the ancients had knowledge only of dead force, and this is what is commonly called mechanics," while it was Galileo who "began to deal with *living force*."[14] He omits that, to the mind of the ancients, the entire universe was living, ensouled, not mechanical but organic or animal, whereas the moderns deal with life as an exception within an overwhelmingly lifeless cosmos. The Aristotelian distinction between *phusis* and *techné*, the self-emergent, plantlike growing whole of nature and the collection of manufactured artifacts that do not contain their origins in themselves, should have given him a pause and made him qualify his statement.

In the same year (1695), Leibniz anonymously published his *Système nouveau de la nature*, where he acknowledged his debt to ancient Greek philosophy, dispelling the illusion that he had earlier freed himself "from the yoke of Aristotle," *je m'étais affranchi du joug d'Aristote*.[15] Besides brushing off the previously cherished anti-Aristotelian idea that a void existed in nature, the anonymous author of *Système* resolved to restore the substantial forms inspired by Aristotle's first entelechies. "I call them, perhaps more intelligibly, *primitive forces*," he noted, "which contain not only *act* or the completion of possibility, but also an original *activity* [*qui ne contiennent pas seulement l'*acte *ou le complément de la possibilité, mais encore une* activité *originale*]."[16] Leibniz will not advance toward energeia any further than this. Remember that entelechy, which we have translated with the neologism *enendment*, completes and guides material potentialities toward their fulfillment. Enendment largely overlaps with energetic enworkment, and, indeed, *Système* imputes to real forms the substantive and verbal senses of work, *activity* and *act*, grouped under the banner of energeia. But in the recovery of entelechy metaphysics likewise prevails over physics: the unities of form—reminiscent of monads—are the "metaphysical points," *points métaphysiques*, that "have *something vital*, a kind of perception," unlike the "physical points," *points physiques*, "indivisible only in appearance."[17] Leibniz's substances are *eo ipso* subjects, unlike matter, which, taken in isolation from form, is hardly real. The unity of matter and form can be assured by nothing other than metaphysical energy, seeing that physical energy, impotent to accomplish the work of synthesis, causes things to fall apart. Were Leibniz to have entertained energy dreams bordering on his flirtations with entelechy, he would have concluded that energy dies, not rests, in matter.

———— ✸ ————

Despite Husserl's critique of Einstein, transiently launched in *The Crisis of European Sciences and Transcendental Phenomenology*, the premises of the theory of relativity resemble those of phenomenology, with which it is contemporaneous. Husserl ventures to bring philosophy back to the things themselves, tracing the abstractions of conceptual and geometrico-mathematical thinking to their experiential foundation in the lifeworld. Einstein, in turn, writes that the "only justification for our concepts and systems of concepts is that they serve to represent the complex of our experiences; beyond this they have no legitimacy."[18] Both phenomenology and relativity physics begin with the updated concepts of time and space, prompting them to rejoin their material, experiential base. The outcomes of the two intellectual exercises are also similar. They prove that, rather than ideal unities, time and space are relative to the affective, cognitive, and other states of the subject (phenomenology) or to the unique and essentially uneven distributions of matter and energy (relativity).

Einstein was alive to the fact that his far-reaching challenge to Newtonian physics fatally interfered with the philosophical a priori ideas of space and time: "I am convinced that the philosophers have had a harmful effect upon the progress of scientific thinking in removing certain fundamental concepts from the domain of empiricism, where they are under our control, to the intangible heights of the *a priori*."[19] To be more precise, the space-time of relativity is not empirically accessible either; it requires an imaginative leap—another way of dreaming than that sanctioned by models of three-dimensional space or clock time—to what experience (of humans and not only) will have been in light of the energy-mass configurations strange and unfamiliar to us. The experience of an object (subjective and objective genitive) on the edge of a black hole does not fit the mold of an empirical given, much less of a transcendental condition of possibility for other experiences. Perhaps without accounting for all the implications of their epistemic revolutions, Einstein and Husserl, who in *Cartesian Meditations* probed the limits of accessing the experience of the other, carved out a niche for thinking and being outside the empirical/transcendental divide.

Compare Einstein's theory of space and time to Kant's. In *Critique of Pure Reason* these are the two ingredients of the "transcendental aesthetic," a precondition for having sensuous experience. But what if, Einstein might ask, these conditions were themselves conditioned? Energy and mass, the two modes of appearing that belong to the same reality, do not occur in preexistent space and time but, depending on their densities, alignments, and distributions, modify the experience of space-time. As Tim Maudlin specifies, "the distribution of matter and energy *constrains* the geometry of space-time but does not *determine* it."[20] Which is to say that it trans-transcendentally outlines the space-time limits, instead of empirically setting them in stone. It is this trans-transcendental domain, which is paradoxically hyperreal, that Einstein and Husserl have discovered.

Matter and energy put to work, activate, actuate, and actualize our experience of space-time. There is no space as such—only "spaces of reference," some of them physically equivalent or inertial, and others not so vis-à-vis each other.[21] Neither time nor length (i.e., temporal and spatial *intervals*) is absolute,[22] given that their delimitations depend absolutely on the singular distributions of matter and energy. When these distributions disobey the physical principle of inertia, we make a transition from the special to the general theory of relativity.[23] All the same, the derivation of coordinates for experience from variations in energy and mass render space and time "physically real," a conclusion that catapults us back to Einstein's initial pronouncement concerning the need for an experiential deduction of ostensibly abstract concepts: "From all of these considerations, space and time data have a physically real, and not a mere fictitious, significance; in particular this holds for all the relations in which co-ordinates and time enter."[24] For the purposes of our study, however, "physical reality" denotes the reciprocal fulfillment—I would like to write *inter-fulfillment*—of energy and space-time.

In a flashback to the plenitude of energeia, the energy-mass of relativity physics does not undergo entropy, which leaves us with the sense of being chronically empty-handed. *Au contraire*, it empties out into the fullness of the world, producing the infrastructures of actuality, the Hegelian Wirklichkeit, starting with the actuality of space-time. The specific quantities of matter and energy affect the space-time curvature and, along with it, the volume of objects or particles contained in it.[25] Shrinking or expanding, curved or (more) flattened out space-time is full, deriving as it does from the distributional relations of matter and energy in the universe. Enworked or emplayed between bodies, of the observer and the observed among others, it is not an infinitely stretched and flat plane, let alone an endless continuum. If space-time is inherently material, it is thereby also energetic, brimming with dynamic variations, yet never descending to the abstract level of pure potentiality. Einsteinian relativity infuses the here-and-now, the moment *and* matter, with the fullness of energeia no longer indexed to a transcendent ideal or an extratemporal, metaphysical reality.

Bombarded by the reminders of energetic fullness in theories of relativity, we might misconstrue the key tenets of quantum physics for
signs of privation. The negative form of many quantum concepts—
Heisenberg's *un*certainty principle and Bohr's *in*determinacy principle;
the *dis*continuity of light's particulate matter discovered by Max Planck;
patterns of *dif*fraction elaborated by Barad—could be misinterpreted as
so many insufficiencies in being itself and in our knowledge of it. Yet,
determinacy, continuity, certainty, identity, and gathering are, far from
the markers of the actual plenitude, the gauges of unrealizable metaphysical ideals. To eschew the chimeras of metaphysics, which have seeped
into scientific inquiries and methodologies, is to get in touch with the
energy of finitude, at first through the via negativa characteristic of the
quantum lexicon.

For the purpose of illustrating my argument, consider the coimbrica-
tion of the observer and the observed in quantum physics. "According
[to] the quantum postulate," Bohr writes, "every observation of atomic
phenomena will just involve an individual process, resulting in an
essential interaction. We cannot therefore speak of independent tools
of measurements."[26] This goes against "our usual description of physical
phenomena . . . based entirely on the idea that the phenomena concerned
may be observed without disturbing them appreciably."[27] It follows from
Bohr's postulate that the workings of quantum reality are not independent from the observer, who is affected by them and who activates largely
imperceptible chains of events in them simply by taking measurements.
For classical physics, the inevitable tainting of an observed phenomenon
by the scientific tools at our disposal is a shortcoming, assessed against
the axiological and ontological background of a "true" and self-enclosed
reality. But, for quantum theory, it signifies our emancipation from the
shackles of ideal methodologies that have been spliced between physics
and *what is* for as long as "the West" has existed. More than that, the
"entanglement" of the observer and the observed attests to the synergy,
the putting-to-work-with, constitutive of experience.

The same goes for the nonideal concurrence of the wave and cor-
puscular notions of light, which, in keeping with Bohr's *principle of
supplementarity*, "are able only to account for complementary sides of

the phenomena."[28] In the universe, the way light is actually enworked or emplayed is complex: it travels as a wave *and* as a particle, in a peculiar synergy of continuity and discontinuity. Qua energy—the working and the work, act and activity, actualizing and actuality—light is a "specific material arrangement" that subsumes the extremes in its movement.[29] There is a boisterous energetic plenitude in the principle of supplementarity, always in excess of the ideal's flat absoluteness. Embracing the opposites, the energy of light abstains from synthesizing them and, instead, enworks or emplays itself as the one and the other, irreducible to each other.

With regard to space-time and its relation to energy, Bohr goes much further than Einstein. In a 1931 lecture, "Space-Time Continuity and Atomic Physics," Bohr suggests that the quantum materiality of space-time entails singular events that cannot be sequentially linked in usual time chains or contiguously assembled in spatial points with determinate empty intervals separating them. Any comparison of events on standard time- or space-scales is built on an unarticulated premise of the neutrality of measurements with respect to the measured, which is not at all the case in quantum theory:

> We have thus either time-space description or description where we can use the laws of conservation of energy and momentum. They are complementary to one another. We cannot use them both at the same time. If we want to use the idea of space-time we must have watches and rods which are outside and independent of the object under consideration, in that sense that we have to disregard the interaction between the object and measuring rod used. We must refrain from determining the amount of momentum that passes into the instrument in order to be able to apply the space-time point of view.[30]

The law of the conservation of energy is, if not outright rejected, unascertainable using time and space measurements, because to confirm it one at least needs to establish the quanta of energy "before" and "after" an event. But since "watches and rods" are inextricably linked to the event they strive to record, assuming that they receive a fraction of its energy ("the amount of momentum that passes") and give some of their energy to it, attempts to verify that the total amount of energy has remained unchanged wind up in a deadlock. We could certainly conjure up a metameasurement that controls the original one and registers its quantum entanglements; this, however, would lead to the problem of a methodological infinite regress, the need to gauge the entanglements of the metameasurement, and so forth.

Not only is the conservation of energy indeterminable and therefore denied the prestige of a principle or a law, but also matter and energy are more tightly bound to each other than they have been in their conversions—that is to say, the operations still tinted with the ideality of equivalence and seeking to establish the exact identity on both sides of the equation—of relativity physics. Although it looks like quantum physics emulates the gesture of formalizing and quantifying matter no longer as mass but as Planck's *quanta of action*, its stress on *action* initiates a daring speculative dialectic of matter and energy. Quantum energy *is* matter in all its instability, singularity, and spontaneous emissions at the atomic and subatomic levels. Matter *is* quantum energy, activated and active above all in its presumed rest, outside the conventional binary of activity and passivity. With more success than the theory of relativity, quantum theory slips away from the nets of the Husserlian criticism of physics, which "whether represented by a Newton or a Planck or an Einstein, or whomever else in the future, was always and remains exact science. It remains such even if, as some think, an absolutely final form of total theory-construction is never to be expected or striven for."[31] Quantum indeterminacy forefends Husserl's attack, to the extent that it receives this quality from matter itself, staying with the things (the matters, stuff, *Sachen*) themselves, so dear to the phenomenological heart. It thus imbibes the energy of matter and imparts it to thought.

———— ∞ ————

Physical fancies are not (only) the fancies of physics. If energy dreams in and as matter, be it organic or inorganic, then physics cannot in good conscience elevate itself to the metonymy of the physical. Often enough, the workings of physical energy contradict or counteract those that are congruent with the energy of physics. At the juncture between the physical and physics, the field of biology finds its proper place.

Avicenna's *Compendium on the Soul* is a forerunner of philosophical approaches attentive to the disparities between physical and biological energies. There, he classifies two kinds of movements: those that proceed according to the element and those that militate against it. The naturalness of the first is congenial to Aristotelian physics, where gravitational energy augurs the return of an object (say, a stone) to the material that predominates in its constitution (in this case, the earth). The other energy breaks this law, concerning as much ontology as cosmic justice, and propels the object away from the element to which it appertains. A physical trajectory disobeying the course preordained to inanimate things is that of living entities, whose physical energy is simultaneously metaphysical or spiritual. For Avicenna, this is an adequate explanation for "a flying bird's motion with its heavy body high up through the sky,"[32] as well as for growth, which directs the earthy and watery body of a plant to the airy realm above it. Perplexing within the mechanical order of the universe, the energy of avian flight and vegetal proliferation are emblematic of the soul that imposes its singular form on matter in each living being, animated thanks to this imposition.

In a certain sense, the branching of physical energy into the energies of physics and biology obviates Husserl's critique of exact sciences, primarily aimed at the former discipline. The scientific question of energy receives a response that is essentially indeterminate, suspended between the "exact" and "life" sciences. Galileo once quipped that "[an Aristotelian] must believe that if a dead cat falls out of a window, a live one cannot possibly fall too, since it is not a proper thing for a corpse to share in qualities that are suitable for the living."[33] But the fact of the matter is that the energy of the living is not entirely compatible with the mechanics of inanimate phenomena, in spite of life's animation by the transformation of chemical energy, through metabolism, into the kinetics of motion or growth. (By the way, Aristotle would deem *kinetic energy* redundant, given that all energy is a function of *kinesis*, i.e., movement in a sense more ample than "change of place.") There is a modicum of freedom in the enworkment of the living that deviates from the closed system of physics. Energy is divided in and against itself. Absent its inherent counterwork or counteraction, its concept will have been incomplete.

Centuries later, Gregory Bateson will reinforce the argument for biological exceptionalism, this time by turning to dogs, rather than to cats, as Galileo did: "When one billiard ball strikes another, there is an energy transfer such that the motion of the second ball is energized by the impact of the first. In communicational systems, on the other hand, the energy of the response is usually provided by the respondent. If I kick a dog, his immediately sequential behavior is energized by his metabolism, not by my kick. . . . He may turn and bite."[34] Physics does not account for what Bateson calls *the energy of the response* surpassing a reaction of objects, which depends on their mass, velocity, elasticity, etc. As matter worked upon by the boot, the dog's body is no different from a billiard ball hit by a cue stick or by another ball. But as the enactment or enworkment of the dog's life, this body moves within the degrees of free-dom permitted by its instinct and environmental experience: upbring-ing, conditioning, learning . . . In addition to the conservation of energy, observed in the game of billiards, or whenever one body strikes another, biological responses let additional energy into the mix, drawing on other founts, *unconnected to the source of the impact and its absorption by the recipient.* The dog biting whoever hit it is a biological counterwork to the putting-into-work of the kick.

Quantum theory promises to marry physics and biology, while doing everything in its power to preserve the indeterminate discrepancy between their respective energies. Barad's moniker for this reconcilia-tion is *material agency*,[35] which is to say, the agency of a dog, a cat, or any other living entity, along with that of a billiard ball, a measuring rod, or any material body imaginable. Bohr, nonetheless, speaks of "the impossibility of carrying out a coherent causal representation of quan-tum phenomena,"[36] the very representation, upon which the concepts of mechanical interactions, agency, freedom, and determinism depend. In this respect, quantum physics recaptures an additional dimension of Aristotle by way of repudiating impoverished effective causality, the cause-effect relation being but one of four Aristotelian categories for a cause. Although he keeps final causes under lock and key in the dusty drawer of intellectual history, Bohr still restitutes to matter, whether "living" or not, its freedom to play or work, to emplay or to enwork itself, outside deterministic constraints. And we, the presumably detached observers, have our share in material freedom, the freedom of matter everywhere putting itself to work or into play.

Beyond a crude juxtaposition of vitalism and mechanism, beyond, also, the age-old extremes of analysis and synthesis, quantum matter-energy always escapes, slips away, sheds the semblance of an identity, and, its withdrawal infinitely passing into absolute givenness, inter-acts with us, as us. Its flight from us, which, at the same time, hands us over to ourselves, is an oblique indication of its (and our) freedom. Neither cause nor effect, it is the in-between where articulation resembles disarticulation and discrete pockets of energy are consistent with entanglements. Energy dreams: on the hinge of freedom.

THE LAST WORD

Energy or Energies?

The question of energy is among the foremost issues of our time, which is why it is all the more surprising that it has not been raised *as* a question, at least not recently. One of my hopes for the book you have just finished reading is that it will rekindle the urge to interrogate energy (and, with it, ourselves, who are inseparable from it) as to its ontological underpinnings governing the ways we treat the world around us, our bodies, and minds. In pursuit of alternatives to our destructive habits of procuring energy, the approach I advocate will be unassuaged by minor adjustments in the sources or modes of extraction applied to this prized object. Nothing short of fundamental changes in the architecture or the infrastructure of energy will do, casting a shadow of doubt, above all, on its purely objective standing. Having embarked on—and having opened ourselves to—such an interrogation, we become cognizant of its inexorable doubling. The question of energy splits into *who?* and *what?*—just as that which is questioned betrays itself as more than one in the one, given its active and actualized, moving and resting, verbal and substantive senses.

Wide-ranging as my analysis of the subject has been, it should not give you the impression that there is only one energy, subsequently shattered into a plurality of expressions, from the economic to the psychological, theological to political, philosophical to physical. Such a conjecture would, indeed, reinstall the concept in the metaphysical lineage to which it is usually relegated. The last word, if there is to be one, is going to be in the plural. As I have repeated several times in this book, energy, inhabited by originary difference, is multiple, heterogeneous, and synergic from the get-go, or, as philosophers say, a priori. Among others, Palamas, Keynes, Freud, Schmitt, Planck, Bohr, and Barad have sent us a legion of reminders to that effect. From them, we learn that *energies* is a plurality that does not boil down to a unified concept, its formulation infinitely postponed by Aristotle for this very reason. But this dissemination, on all fronts of human existence, will not thwart our enacting and enthinking of energetic ontology. On the contrary, as I have tried to convey in *Energy Dreams*, our mindful relation to energies will be reinvigorated in light of their extraconceptual proliferation.

The work is cut out for us, and that work, too, is the enworkment of energy. We must hone, in our minds and our hands, the skill of holding together, without either prioritizing or denigrating them, the energetic actuality of the ancients and the energetic potentiality of the moderns. In the process, which is by the same token the outcome, we will appreciate how to rest with, or in, movement, and to move with, or in, rest; to cherish the surface, despite our fascination with depth; to let energies blossom in their sharing that reaffirms their originary differences; to embrace finitude, rather than demolish the world in the name and on the basis of our fantasy of indestructability.

It would be wise to take the mention of fantasy in regard to our harmful treatment of existence as a hint that the work to come involves dreams, among other things. The new enworkment of energy cannot get off the ground without dreamwork, an analysis of our motley dreams about energy and an exercise in imagining what another energy, or other energies, independent of the default extractive-destructive paradigm, would look like. Here is to energetic dreaming, shimmering with the hidden potentialities of the actual and the effective actuality of the potential!

P.S.—THE VERY LAST WORD

Y ou are welcome to read the title of this work in a variety of ways that jointly maintain the fruitful ambiguities of energy and resist the urge to resolve its crisis in thought alone. First, energy is a subject who dreams, while the object it dreams about is actuality. Since, in an important sense, energy *is* actuality, it ends up dreaming about itself. Second, whatever occupies the center of my theoretical attention here (for instance, the dreams energy has, or dreams to do with energy), the book is also, by the same token, a treatise on actuality. *Of* enters into a relation of semantic equivalence with *on*. Hence, "Of Actuality" or "On Actuality," following the Latin (and medieval) manner of bestowing titles onto philosophical works, among which we find Lucretius's *De rerum natura*, "On the Nature of Things," Duns Scotus's *De rerum principio*, "Of the Beginning of Things," or Aristotle's *De anima*, "On the Soul," in keeping with the standard rendition of the Greek *Peri psukhēs*. Third, energy dreams—grammatical form indeterminate—pertain to actuality. They are *of* actuality: partaking of, participating in, proper to, or appropriated by the actual, including the mundane level of *l'actualité* as "current events," the news cycle.

Actuality is energy's very last, but also very first, word. It makes possible the consonance of the beginning and the end. Those of metaphysics, above all. We find it more and more difficult, however, to hear the distinct overtones of this word and, provided that we *do* hear something, to listen to what it expresses. To attune our ears to it, we must learn to discern vibrations so minute that they no longer register within our auditory range. I'd like to believe that the book you are about to close could prepare the grounds for this discernment by serving as a tuning fork for energy-dreams-actuality. That, at least, is how I am dreaming of its future.

NOTES

OPENING WORDS

1. Cf. Marder, *Pyropolitics*.

1. ENERGY DREAMS

1. Cf. Bibikhin, *Energiya*, 13.
2. Urmson, *The Greek Philosophical Vocabulary*, 61.
3. Winters, "Lexical Layers," 450.
4. Heidegger, *The Event*, 87; see, especially, Heidegger's lecture course on *Aristotle's* Metaphysics Theta *1-3*.
5. Derrida, *Writing and Difference*, 279–80.
6. In distinct ways and through different routes, I owe this reading to Giorgio Agamben (*impotentiality*) and Vladimir Bibikhin (*energiya*).
7. For more on this classification, refer to chapter 5 of the present study.
8. Heidegger, *Being and Time*, 189.
9. Losev, *Ocherki*, 731–32.
10. Hegel, *Phenomenology of Spirit*, 359.
11. Hegel, *Philosophy of Right*, 10.
12. Ibid., 13.
13. Hegel, *Lectures on the History of Philosophy*, 138.
14. Hegel, *Science of Logic*, 485.
15. Ibid., 486.
16. Hegel, *Philosophy of Right*, 11.
17. Hegel, *The Encyclopaedia Logic*, 215.

18. Hegel, *Lectures on the Philosophy of World History*, 50 (translation modified).
19. For more on the work of the thing, see my *The Event of the Thing*.
20. Heidegger, *Being and Time*, 79.
21. Heidegger, *Basic Writings*, 165.
22. Ibid., 168.
23. Ibid., 165.
24. Heidegger, *The Basic Problems of Phenomenology*, 105.
25. Ibid., 101.
26. Ibid., 106.

2. THEOLOGICAL MUSINGS

1. Vattimo, *Belief*, 39.
2. Agamben, *The Highest Poverty*, 123.
3. Nietzsche, *On the Genealogy of Morality*, 63.
4. Ibid., 20.
5. Nietzsche, *The Will to Power*, 340.
6. Ibid.
7. Beaufret, *Dialogue with Heidegger*, 112.
8. Eckhart, "Sermon XXV," 218.
9. Eckhart, *Et Ce Néant Était Dieu* . . . , 103 (translation mine).
10. Eckhart, *Selected Writings*, 234 (translation lightly modified).
11. In the pre-Hispanic, and hence pre-Christian, societies of the Andes, a synergetic theology thrived under the title that was later to be translated as *crianza mútua*, "mutual creation." It entailed equality among co-created living beings, neither of whom could usurp for itself the place of the Creator. The two principles of *crianza mútua*—"I create to be created" and "As I create, I am created"—establish a radical synergy of working-with-the-other. I thank Eduardo Molinari for this precious reference.
12. Thoreau, *Walden*, 76.
13. Cf. Horujy, "Synergia as a Universal Paradigm."
14. Luce Irigaray has recently indicated one possible approach to thinking the energy of yoga in her *Toward a New Culture of Energy*.
15. For other points of agreement between yoga and Hesychasm, consult Oommen's *A Christian Outlook on Yoga*, 66ff.

3. ECONOMIC CHIMERAS

1. For more on *nomos* as the inner measure, see Schmitt, *The Nomos of the Earth*.
2. Derrida, *Dissemination*, 87.
3. Marx, *Capital*, 1:117–18.
4. Keynes, *The General Theory*, 167.

5.   Ibid., 222.

6.   Howells, "Velocity and the Money Multiplier," 791.

7.   Keynes, *The General Theory*, 194. Cf., also, Bibow, *Keynes on Monetary Policy*, 29.

8.   Mill, *The Principles of Political Economy*, 22.

9.   Keynes, *The General Theory*, 129.

10. Mill, *The Principles of Political Economy*, 6 (emphasis added).

11. Marx, "Economic and Philosophic Manuscripts of 1844," 71.

12. Ibid., 73–74.

13. Ibid., 74.

14. Ibid., 99.

15. Karl Marx, *Capital*, 1:1044.

16. Ibid.

17. Bataille, *The Accursed Share I*, 19–20.

18. Smith, *The Wealth of Nations*, 14.

19. Ricardo, *The Principles of Political Economy and Taxation*, 81. I owe this reference to Patrícia Vieira.

20. Friedman, *Capitalism and Freedom*, 15.

21. Ibid., 59.

22. Joseph Townsend's 1786 *A Dissertation on the Poor Laws* made no secret of the "peaceable, silent, unremitting pressure" exerted by the threat of starvation that acted "as the most powerful natural motive to industry and labor. "Hunger will tame the fiercest animals, it will teach decency and civility, obedience and subjection, to the most perverse. In general it is only hunger which can spur and goad them [the poor] on to labor; yet our laws have said they shall never hunger. . . . Legal constraint is attended with much trouble, violence, and noise . . . whereas hunger is not only peaceable, silent, unremitting pressure, but as the most powerful natural motive to industry and labor, it calls for the most powerful exertions." Quoted in Polanyi, *The Great Transformation*, 113.

23. For more on the sense of dwindling energies in modernity, refer to Bibikhin, *Energiya*, 59ff.

24. Demirel, *Energy*, 175.

25. Albritton, *Let Them Eat Junk*, 151.

26. Weis, *The Global Food Economy*, 42.

27. Lorigan, *The Rise and Fall of High Fructose Corn Syrup and Fibromyaglia*, 77ff.

4. PSYCHOLOGICAL REVERIES

1.   "By the nutritive faculty I mean that part of the soul, which even the plants participate in [θρεπτικόν δέ λέγομεν τὸ τοιοῦτον μόριον τῆς ψυχῆς οὐ καὶ φυτὰ μετέχει]" (*De anima* 413b, 7–8).

2.   Here and thereafter, *SE* refers to the standard edition of Freud's writings.

3.   Jung, *Psychological Types*, 262.

4.   Ibid.

5.   Ibid.

6.   Ibid., 571.

7.   Lacan, *The Seminar of Jacques Lacan*, bk. 2, 74, 75.

8.   Lacan, *Écrits*, 550.

9.   Lacan, *Seminaire XIV*, Session 22/2/67, 147.

10. Ibid.

11. Lacan, *Écrits*, 68.

12. Lacan, *The Seminar of Jacques Lacan*, bk. 2, 305.

13. In addition, Lacan's generalization of "natural phenomena" overlooks significant incongruities between biological and physical energies, to be discussed in chapter 6 (e.g., the way upward vegetal growth overcomes the force of gravity).

14. Lacan, *The Seminar of Jacques Lacan*, bk. 3, 146.

15. This is not to deny that the imaginary also entails a synthetic moment of identification across the void of alienation, the moment indispensable for self-recognition and for the integration of one's body image.

16. Lacan, *Écrits*, 552.

17. Ibid., 95.

18. Lacan, *The Seminar of Jacques Lacan*, bk. 2, 257.

5. POLITICAL PHANTASIES

1.   Kant, "Perpetual Peace," 93 (translation modified).

2.   Ibid., 130 (translation modified).

3.   I have made a similar argument earlier when I revisited the Aristotelian distinction between *peras* and *telos*.

4.   Heidegger, *Being and Time*, 288–89.

5.   This text was originally given as a talk at an international conference on "the end of the world" in Paris, France, September 2015.

6.   Marder, "Philosophy's Homecoming."

7.   Schmitt, *The Nomos of the Earth*, 59–60.

8.   The concepts of actuality and activity correspond, of course, to the substantive and verbal senses of *ergon*, work.

9.   Heidegger, *Being and Time*, 63.

10. Ibid., 436.

11. Sorel, *Reflections on Violence*, 73.

12. Ibid., 2.

13. Ibid., 165–66.

14. Ibid., 28–29.

15. Benjamin, "Critique of Violence," 291–92.

16. Sorel, *Reflections on Violence*, 77.

17. Bibikhin, *Energiya*, 315 (translation mine).

18. Ugilt, *The Metaphysics of Terror*, 6.

19. Benjamin, "Critique of Violence," 292.
20. Hegel, *Philosophy of Right*, 155.
21. Ibid. (translation modified).
22. Ibid., 160.
23. Hegel, *Phenomenology of Spirit*, 310.
24. Ibid.
25. Magun, *Negative Revolution*, 156ff.
26. Le Bon, *The Crowd*, 2.
27. Le Bon, *The Evolution of Forces*, 10.
28. Ibid., 68.
29. Schmitt, *The Concept of the Political*, 38.
30. Ibid.
31. Schmitt, *Political Theology*, 5.
32. Ibid., 15.
33. Quoted ibid. (translation modified).
34. Schmitt, *Constitutional Theory*, 67.
35. Ibid., 59, 61.
36. Ibid., 61.
37. Schmitt, *The Crisis of Parliamentary Democracy*, 26.
38. Ibid., 6.
39. Schmitt, *Roman Catholicism and Political Form*, 34.

6. PHYSICAL FANCIES

1.   Alexievich, "About Loving Physics," 179–84 (translation mine).
2.   Vincent, *Prisoners of Hope*, 32.
3.   Barad, *Meeting the Universe Halfway*, 124.
4.   For an excellent foray into related questions with regard to the second law of thermodynamics, cf. Schrader, "Haunted Measurements," 119–60.
5.   Smolin, *Time Reborn*, 241.
6.   Compare this to Einstein's "the energy, E0, of a body at rest is equal to its mass." Einstein, *The Meaning of Relativity*, 47.
7.   Barad, *Meeting the Universe Halfway*, 236.
8.   Ibid., 152–53.
9.   Hegel, *Philosophy of Nature*, 52.
10. Ibid., 45–46.
11. Leibniz, *Philosophical Essays*, 118.
12. Kavanaugh, *The Architectonic of Philosophy*, 192.
13. Feldman, "Energy," 122.
14. Leibniz, *Philosophical Essays*, 122.
15. Ibid., 139.
16. Ibid.

17.   Ibid., 142.
18.   Einstein, *The Meaning of Relativity*, 2.
19.   Ibid., 2.
20.   Maudlin, *Philosophy of Physics*, 139.
21.   Einstein, *The Meaning of Relativity*, 24.
22.   Ibid., 25–26.
23.   Ibid., 60ff.
24.   Ibid., 29.
25.   "The way the volume of small balls of test particles in free fall behave in a region is determined by the amount of matter and energy in that region. The more matter and energy, the greater the Einstein curvature and the more the volume of the balls will shrink." Maudlin, *Philosophy of Physics*, 139.
26.   Bohr, *Collected Works*, 6:75; see also Bohr, *Collected Works*, 10:41.
27.   Ibid., 6:148.
28.   Ibid., 6:76.
29.   Barad, *Meeting the Universe Halfway*, 296.
30.   Bohr, *Collected Works*, 6:369.
31.   Husserl, *The Crisis of European Sciences and Transcendental Phenomenology*, 4.
32.   Avicenna, *A Compendium on the Soul*, 23.
33.   Quoted in Coopersmith, *Energy, the Sutble Concept*, 15.
34.   Bateson, *Steps to an Ecology of Mind*, 229, 409.
35.   Barad, *Meeting the Universe Halfway*, 34.
36.   Bohr, *Collected Works*, 6:285.

BIBLIOGRAPHY

Agamben, Giorgio. *The Highest Poverty: Monastic Rules and Form-of-Life.* Trans. Adam Kotsko. Stanford: Stanford University Press, 2013.

Albritton, Robert. *Let Them Eat Junk: How Capitalism Creates Hunger and Obesity.* New York: Pluto, 2009.

Alexievich, Svetlana. "About Loving Physics." In *Voices from Chernobyl.* Trans. Keith Gessen London: Dalkey Archive, 2005.

Aquinas, St. Thomas. *Summa Theologiae,* Latin and English ed. 8 vols. Lander, WY: Aquinas Institute, 2012.

Aristotle. *Metaphysics. Oeconomica. Magna Moralia.* Loeb Classical Library vols. 271 and 287. Cambridge: Harvard University Press, 1933–1935.

——. *Nicomachean Ethics.* Loeb Classical Library vol. 19. 2d ed. Cambridge: Harvard University Press, 1934.

——. *On the Soul. Parva Naturalia. On Breath.* Loeb Classical Library vol. 288. Rev. ed. Cambridge: Harvard University Press, 1975.

——. *The Physics.* Loeb Classical Library vols. 228 and 255. Rev. ed. Cambridge: Harvard University Press, 1934–1957.

——. *The Politics.* Loeb Classical Library vol. 264. Cambridge: Harvard University Press, 1932.

Augustine, St. *Confessions.* Oxford: Oxford University Press, 2009.

——. *Epistula: Letters,* vol. 2. Trans. Sister Wilfrid Parsons. Washington, DC: Catholic University of America Press, 1951.

——. *In Ioannis Evangelium Tractatus: Homilies on the Gospel of John.* Trans. John Gibb. In Paul Schaff, ed., *Nicene and Post-Nicene Fathers,* vol. 7. New York: T&T Clark, 1886.

——. *The City of God, Against the Pagans.* Trans. R. W. Dyson. Cambridge: Cambridge University Press, 1998.

Avicenna. *A Compendium on the Soul.* Trans. Edward Abbott van Dyck. Verona: Stamperia di Nicola Paderno, 1906.

Barad, Karen. *Meeting the Universe Halfway: Quantum Physics and the Entanglement of Matter and Meaning*. Durham: Duke University Press, 2007.

Bataille, Georges. *The Accursed Share: An Essay on General Economy*, vol. 1: *Consumption*. Trans. Robert Hurley. New York: Zone, 1991.

Bateson, Gregory. *Steps to an Ecology of Mind: Collected Essays in Anthropology, Psychiatry, Evolution, and Epistemology*. Chicago: University of Chicago Press, 2000.

Beaufret, Jean. *Dialogue with Heidegger: Greek Philosophy*. Trans. Mark Sinclair. Bloomington: Indiana University Press, 2006.

Benjamin, Walter. "Critique of Violence." In *Reflections: Essays, Aphorisms, Autobiographical Writings*. Trans. Edmund Jephcott. New York: Schocken, 1978.

Bibikhin, Vladimir. *Energiya* [Energy]. Moscow: St. Thomas Institute of Philosophy, Theology and History, 2010.

Bibow, Jörg. *Keynes on Monetary Policy, Finance and Uncertainty: Liquidity Preference Theory and the Global Financial Crisis*. New York: Routledge, 2009.

Bohr, Niels. *Collected Works*, vol. 6: *Foundations of Quantum Physics I (1926–1932)*. Ed. Jorgen Kalckar. New York: North-Holland, 1985.

——. *Collected Works*, vol. 10: *Complementarity Beyond Physics*. Ed. David Favrholdt. New York: Elsevier, 1999.

Coopersmith, Jennifer. *Energy, the Subtle Concept: The Discovery of Feynman's Blocks from Leibniz to Einstein*. Oxford: Oxford University Press, 2010.

Demirel, Yaşar. *Energy: Production, Conversion, Storage, Conservation, and Coupling*. London: Springer, 2012.

Derrida, Jacques. *Dissemination*. Trans. Barbara Johnson. London: Continuum, 2004.

——. *Writing and Difference*. Trans. Alan Bass. Chicago: University of Chicago Press, 1978.

Eckhart, Meister. *Et Ce Néant Était Dieu . . . Sermons LXI à XC*. Paris: Albin Michel, 2000.

——. *Selected Writings*. Ed. Oliver Davies. New York: Penguin, 1994.

——. "Sermon XXV." In *Meister Eckhart: Teacher and Preacher*. Ed. Bernard McGinn and Frank J. Tobin. Mahwah, NJ: Paulist, 1986.

Einstein, Albert. *The Meaning of Relativity*. Trans. Edwin Plimpton Adams. New York: Routledge, 2003.

Feldman, Theodore S. "Energy." In William F. Bynum, E. Janet Browne, and Roy Porter, eds., *Dictionary of the History of Science*. Princeton: Princeton University Press, 1981.

Freud, Sigmund. *The Standard Edition of the Complete Psychological Works of Sigmund Freud*. Ed. and trans. James Strachey. 24 vols. London: Vintage, 2001.

Friedman, Milton. *Capitalism and Freedom*. 40th anniversary ed. Chicago: University of Chicago Press, 2002.

Hegel, G. W. F. *The Encyclopaedia Logic*. Trans. T. F. Geraets, W. A. Suchting, and H. S. Harris. Indianapolis: Hackett, 1991.

——. *Lectures on the History of Philosophy*. Trans. E. S. Haldane and F. H. Simson. 3 vols. New York, 1892.

——. *Lectures on the Philosophy of World History*. Trans. H. B. Nisbet. Cambridge: Cambridge University Press, 1981.

——. *Phenomenology of Spirit*. Trans. A. V. Miller. Oxford: Oxford University Press, 1977.

——. *Philosophy of Nature: Encyclopedia of the Philosophical Sciences, Part II.* Trans. A. V. Miller. Oxford: Oxford University Press, 2004.

——. *Philosophy of Right.* Trans., with notes, T. M. Knox. Oxford: Oxford University Press, 1967.

——. *Science of Logic*, vol. 2. Trans. W. H. Johnston and L. G. Struthers. London: MacMillan, 1929.

Heidegger, Martin. *Aristotle's Metaphysics Theta 1–3: On the Essence and Actuality of Force.* Trans. Walter Brogan and Peter Warnek. Bloomington: Indiana University Press, 1995.

——. *The Basic Problems of Phenomenology.* Trans. Albert Hofstadter. Rev. ed. Bloomington: Indiana University Press, 1988.

——. *Basic Writings.* Ed. David Farrell Krell. San Francisco: HarperCollins, 1993.

——. *Being and Time.* Trans. John Macquarrie and Edward Robinson. San Francisco: Harper and Row, 1962.

——. *The Event.* Trans. Richard Rojcewicz. Bloomington: Indiana University Press, 2013.

Horujy, Sergey. "Synergia as a Universal Paradigm: Its Meaning(s), Discursive Links, and Heuristic Resources." International Workshop on *Synergie: Konzepte – Techniken – Perspektiven*, Berlin, June 2011. http://synergia-isa.ru/wp-content/uploads/2011/07/hor_berl_syn_talk.pdf.

Howells, Peter. "Velocity and the Money Multiplier." In Phillip O'Hara, ed., *Encyclopedia of Political Economy*, vol. 2. New York: Routledge, 1999.

Husserl, Edmund. *The Crisis of European Sciences and Transcendental Phenomenology.* Trans. David Carr. Evanston: Northwestern University Press, 1970.

Irigaray, Luce. *Toward a New Culture of Energy.* New York: Columbia University Press, forthcoming.

Jung, Carl. *Psychological Types, or the Psychology of Individuation.* New York: Harcourt, Brace, 1926.

Kant, Immanuel. "Perpetual Peace: A Philosophical Sketch." In H. Reiss, ed., *Political Writings.* Cambridge: Cambridge University Press, 1970.

Kavanaugh, Leslie Jaye. *The Architectonic of Philosophy: Plato, Aristotle, Leibniz.* Amsterdam: Amsterdam University Press, 2007.

Keynes, John Maynard. *The General Theory of Employment, Interest, and Money.* New York: Harcourt, Brace, 1935.

Lacan, Jacques. *Écrits: The First Complete Edition in English.* Trans. Bruce Fink. New York: Norton, 2007.

——. *Seminaire XIV: La Logique du Fantasme, 1966–67.* Session 22/2/67. Paris: Association lacanienne internationale, 2004.

——. *The Seminar of Jacques Lacan*, bk. 2: *The Ego in Freud's Theory and in the Technique of Psychoanalysis, 1954–55.* Ed. Jacques-Alain Miller. New York: Norton, 1991.

——. *The Seminar of Jacques Lacan*, bk. 3: *The Psychoses, 1955–56.* Ed. Jacques-Alain Miller. Trans. Russell Grigg. New York: Norton, 1997.

Le Bon, Gustave. *The Crowd: A Study of the Popular Mind.* New York: Macmillan, 1896.

——. *The Evolution of Forces.* New York: Appleton, 1908.

Leibniz, G. W. *Philosophical Essays.* Ed. Roger Ariew and Daniel Garber. Indianapolis: Hackett, 1989.

Lorigan, Janice. *The Rise and Fall of High Fructose Corn Syrup and Fibromyaglia: Ending Fibromyaglia Without Drugs or Violence.* Bloomington: AuthorHouse, 2011.

Losev, A. F. *Ocherki Antichnogo Simbolizma i Mifologii.* Moscow: Mysl', 1993.

Magun, Artemy. *Negative Revolution: Modern Political Subject and Its Fate After the Cold War.* New York: Bloomsbury, 2013.

Marder, Michael. *The Event of the Thing: Derrida's Post-Deconstructive Realism.* Toronto: University of Toronto Press, 2009.

——. "Philosophy's Homecoming." In Richard Polt and Jon Wittrock, eds., *The Task of Philosophy in the Anthropocene: Axial Echoes in Global Space.* London: Rowman and Littlefield International, forthcoming.

——. *Pyropolitics: When the World Is Ablaze.* London: Rowman and Littlefield International, 2015.

Marx, Karl. *Capital: A Critique of Political Economy,* vol. 1. Trans. Ben Fowkes. New York: Penguin, 1976.

——. *Capital: A Critique of Political Economy,* vol. 2. Trans. David Fernbach. New York: Penguin, 1993.

——. "Economic and Philosophic Manuscripts of 1844." In Robert Tucker, ed., *The Marx-Engels Reader.* New York: Norton, 1978.

Maudlin, Tim. *Philosophy of Physics: Space and Time.* Princeton: Princeton University Press, 2012.

Maximus the Confessor, St. "Opuscula Theologica et Polemica." In J. P. Migne, ed., *Patrologiae Cursus Completus: Patrologiae Graeca,* vol. 91. Paris: Garnier Fratres, 1863.

Mill, John Stuart. *The Principles of Political Economy with Some of Their Applications to Social Philosophy.* Ed. William Ashley. Fairfield: NJ: Augustus M. Kelly, 1976.

Nietzsche, Friedrich. *On the Genealogy of Morality.* Trans. Maudemarie Clark and Alan Swensen. Indianapolis: Hackett, 1998.

——. *The Will to Power.* Trans. Walter Kaufman and R. J. Hollingdale. New York: Vintage, 1968.

Oommen, Abraham. *A Christian Outlook on Yoga.* New Delhi: ISPCK/MPCCET, 2008.

Palamas, Gregory. *The Triads.* New York: Paulist, 1983.

Patañjali. *Enlightened Living: A New Interpretive Translation of the Yoga Sutra of Maharsi Patañjali.* New Delhi: Motilal Banarsidass, 1978.

Plato. *Euthyphro; Apology; Crito; Phaedo; Phaedrus.* Loeb Classical Library vol. 36. Cambridge: Harvard University Press, 1914.

——. *The Republic.* Loeb Classical Library vols. 237 and 276. Cambridge: Harvard University Press, 1930–1935.

Plotinus. *The Enneads.* Trans. A. Armstrong. Loeb Classical Library vol. 441. 7 vols. Cambridge: Harvard University Press, 1966–1988.

Polanyi, Karl. *The Great Transformation.* Boston: Beacon, 1957.

Ricardo, David. *The Principles of Political Economy and Taxation.* London: J. M. Dent, 1911.

Schmitt, Carl. *The Concept of the Political.* Trans. G. Schwab. Chicago: University of Chicago Press, 1996.

——. *Constitutional Theory.* Trans. J. Seitzer. Durham: Duke University Press, 2008.

———. *The Crisis of Parliamentary Democracy.* Trans. Ellen Kennedy. Cambridge: MIT Press, 1986.

———. *The* Nomos *of the Earth in the International Law of* Jus Publicum Europaeum. Trans. Gary L. Ulmen. New York: Telos, 2003.

———. *Political Theology: Four Chapters on the Concept of Sovereignty.* Trans. G. Schwab. Cambridge: MIT Press, 1985.

———. *Roman Catholicism and Political Form.* Trans. Gary L. Ulmen. Westport, CT: Greenwood, 1996.

Schrader, Astrid. "Haunted Measurements: Demonic Work and Time in Experimentations." *differences* 23, no. 3 (2012): 119–160.

Smith, Adam. *The Wealth of Nations* New York: Modern Library, 1937.

Smolin, Lee. *Time Reborn: From the Crisis in Physics to the Future of the Universe.* New York: Houghton Mifflin Harcourt, 2013.

Sorel, Georges. *Reflections on Violence.* Ed. Jeremy Jennings. Cambridge: Cambridge University Press, 1999.

Thoreau, Henry D. *Walden.* Ed. Jeffrey S. Cramer. New Haven: Yale University Press, 2006.

Ugilt, Rasmus. *The Metaphysics of Terror: The Incoherent System of Contemporary Politics.* New York: Bloomsbury, 2012.

Urmson, J. O. *The Greek Philosophical Vocabulary.* London: Duckworth, 1990.

Vattimo, Gianni. *Belief.* Trans. Luca D'Isanto and David Webb. Stanford: Stanford University Press, 1999.

Vincent, Lanny. *Prisoners of Hope: How Engineers and Others Get Lift for Innovating.* Bloomington, IN: WestBow, 2011.

Weis, Anthony John. *The Global Food Economy: The Battle for the Future of Farming.* New York: Zed, 2007.

Winters, Margaret E. "Lexical Layers." In John R. Taylor, ed., *The Oxford Handbook of the Word.* Oxford: Oxford University Press, 2015.

Xenophon. *Oeconomicus.* Trans. E. C. Marchant and O. J. Todd. Loeb Classical Library vol. 168. Cambridge: Harvard University Press, 2013.

INDEX